MATHS PLUS

MENTALS AND HOMEWORK BOOK

AUSTRALIAN CURRICULUM

Harry O'Brien
Greg Purcell

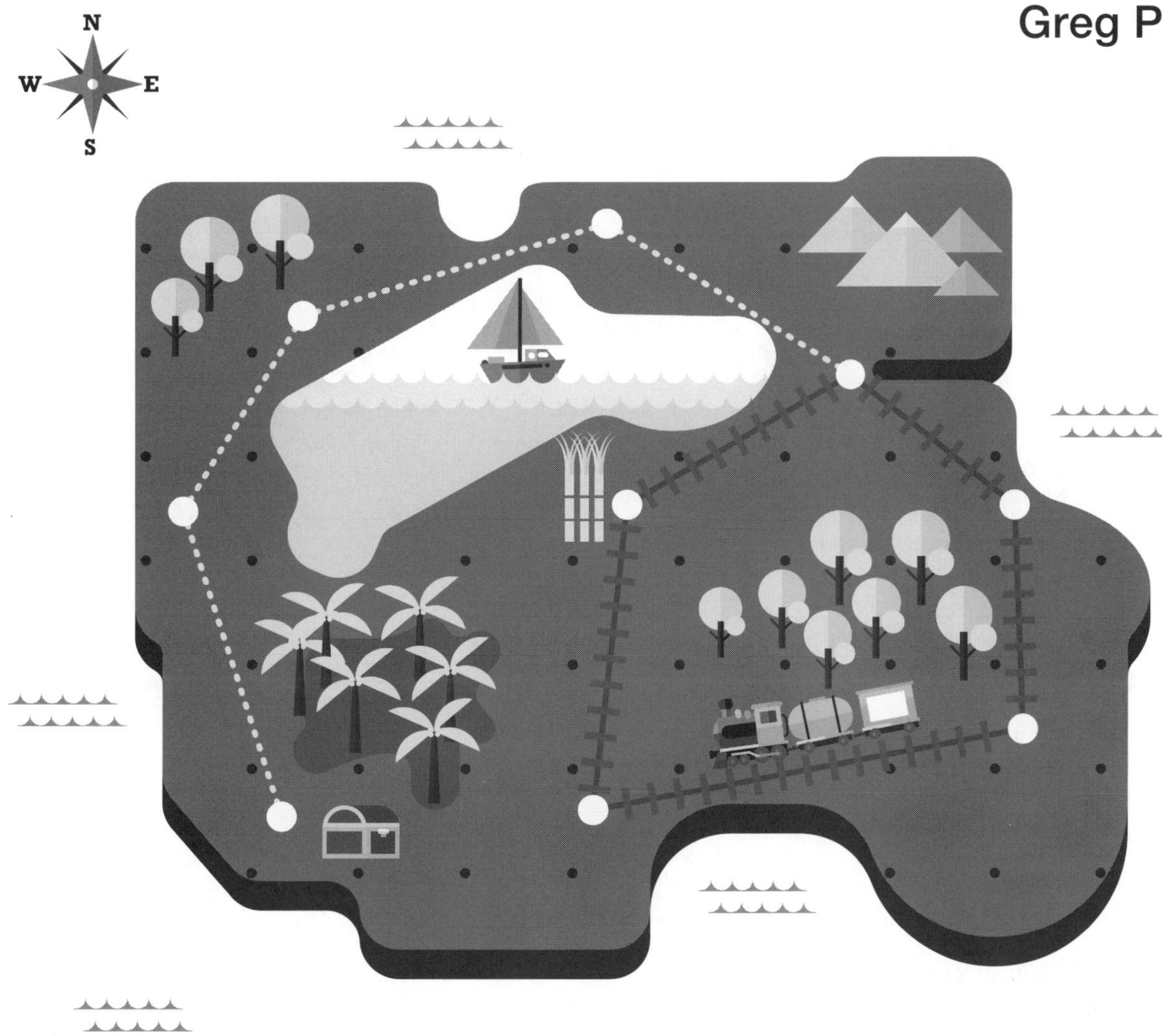

OXFORD
UNIVERSITY PRESS

Contents

Unit	Pages	Number and Algebra		
1	2–3	Place value	Addition and subtraction strategies	
2	4–5	Square numbers	Negative numbers	
3	6–7	Factors	Percentages, fractions and decimals	
4	8–9	Division	Number patterns	
5	10–11	Multiplication strategies	Subtracting decimals	
6	12–13	Negative numbers	Improper fractions and mixed numbers	
7	14–15	Add and subtract fractions	Decimal place value	Decimal number patterns
8	16–17	Problem solving/large numbers	Triangular numbers	
9	18–19	Division	Prime and composite numbers	
10	20–21	Addition and subtraction of 4- and 5-digit numbers	Add and subtract fractions	
11	22–23	Negative numbers	Unit fractions of a quantity	
12	24–25	Order of operations	Percentages, decimals and fractions	
13	26–27	Subtracting decimals	Expanding numbers	
14	28–29	4-digit x 1-digit multiplication	Equivalent fractions	Prime or composite numbers
15	30–31	Comparing and ordering fractions	Rounding	
16	32–33	Division with fractional remainders	Adding fractions—related denominators	Geometric patterns
17	34–35	Extended multiplication	Number patterns	Adding and subtracting decimals
18	36–37	Fractions of a quantity	Applying addition to solve problems	
19	38–39	Operations with decimals	Decimal multiplication/place value	
20	40–41	Multiplication is commutative	Money problems	
21	42–43	Spreadsheets	Equivalent fractions	
22	44–45	Percentages	Geometric patterns	
23	46–47	Multiplying by tens	Rounding money	
24	48–49	Multiplying decimals and money	Number patterns	Finding percentages
25	50–51	Related denominators	Decimal number patterns	
26	52–53	Use factors to solve division	Division of decimals	
27	54–55	Decimals x powers of ten	Spreadsheets	Graphing patterns
28	56–57	Estimation strategies	Dividing decimals	
29	58–59	Problem solving	Solving equations	
30	60–61	Positive and negative numbers	Prime factors	
31	62–63	Negative numbers	Number patterns	
32	64–65	Multiplication problems	Balance	
33	66–67	Decimals x powers of ten	Decimal/fraction number patterns	
34	68–69	Fraction and decimal remainders	Using a calculator	Order of operations
35	70–71	Recurring decimals	Order of operations	

Contents

Unit	Pages	Statistics and Probability	Measurement and Space	
1	2–3	Picture graphs		Angles
2	4–5		Perimeter	Polygons
3	6–7		Square centimetres	Classifying three-dimensional objects
4	8–9	Tree diagram	The cubic centimetre	
5	10–11	Probability from 0 to 1		Rotational symmetry
6	12–13	Column graphs	Timetables	
7	14–15			Representing three-dimensional objects
8	16–17	Line graphs		Constructing angles
9	18–19	Chance/graphs		
10	20–21		Kilometres	Constructing rectangles
11	22–23	Line graphs	Kilograms and tonnes	
12	24–25	Likelihood		Cross sections
13	26–27	Potentially misleading data	Kilometres	
14	28–29			Reflections
15	30–31	Chance from 0 to 1		Problem solving using scale
16	32–33		Calculating volume	
17	34–35		Tonnes and kilograms	
18	36–37		Cubic centimetres and millilitres	Translating shapes
19	38–39	Graphing ordered pairs	24-hour time	
20	40–41	Side-by-side column graphs		Cross sections
21	42–43		Area	Metres, centimetres and millimetres
22	44–45	Graphs	Area and perimeter	
23	46–47	Dot plots/pie charts		
24	48–49		Time	
25	50–51		Grid references	Tessellations
26	52–53	Chance	Measurement units	
27	54–55			Angles
28	56–57	Frequency tables		The Cartesian plane
29	58–59	Chance predictions		Angles
30	60–61	Graphs	Vertically opposite angles	
31	62–63		Capacity units	Tessellating shapes
32	64–65	Interpreting data		The Cartesian plane
33	66–67	Two-way tables	Square and cubic metres	
34	68–69			Making a map
35	70–71	Data exploration	Mass	

Australian Curriculum AC Linking Chart

Units	1	2	3	4	5
Number and Algebra					
Recognise situations, including financial contexts, that use integers; locate and represent integers on a number line and as coordinates on the Cartesian plane (AC9M6N01)					
Identify and describe the properties of prime, composite and square numbers and use these properties to solve problems and simplify calculations (AC9M6N02)					
Apply knowledge of equivalence to compare, order and represent common fractions including halves, thirds and quarters on the same number line and justify their order (AC9M6N03)					
Apply knowledge of place value to add and subtract decimals, using digital tools where appropriate; use estimation and rounding to check the reasonableness of answers (AC9M6N04)					
Solve problems involving addition and subtraction of fractions using knowledge of equivalent fractions (AC9M6N05)					
Multiply and divide decimals by multiples of powers of 10 without a calculator, applying knowledge of place value and proficiency with multiplication facts; using estimation and rounding to check the reasonableness of answers (AC9M6N06)					
Solve problems that require finding a familiar fraction, decimal or percentage of a quantity, including percentage discounts, choosing efficient calculation strategies and using digital tools where appropriate (AC9M6N07)					
Approximate numerical solutions to problems involving rational numbers and percentages, including financial contexts, using appropriate estimation strategies (AC9M6N08)					
Use mathematical modelling to solve practical problems involving natural and rational numbers and percentages, including in financial contexts; formulate the problems, choosing operations and efficient calculation strategies, and using digital tools where appropriate; interpret and communicate solutions in terms of the situation, justifying the choices made (AC9M6N09)					
Recognise and use rules that generate visually growing patterns and number patterns involving rational numbers (AC9M6A01)					
Find unknown values in numerical equations involving brackets and combinations of arithmetic operations, using the properties of numbers and operations (AC9M6A02)					
Create and use algorithms involving a sequence of steps and decisions that use rules to generate sets of numbers; identify, interpret and explain emerging patterns (AC9M6A03)					
Measurement and Space					
Convert between common metric units of length, mass and capacity; choose and use decimal representations of metric measurements relevant to the context of a problem (AC9M6M01)					
Establish the formula for the area of a rectangle and use it to solve practical problems (AC9M6M02)					
Interpret and use timetables and itineraries to plan activities and determine the duration of events and journeys (AC9M6M03)					
Identify the relationships between angles on a straight line, angles at a point and vertically opposite angles; use these to determine unknown angles, communicating reasoning (AC9M6M04)					
Compare the parallel cross-sections of objects and recognise their relationships to right prisms (AC9M6SP01)					
Locate points in the 4 quadrants of a Cartesian plane; describe changes to the coordinates when a point is moved to a different position in the plane (AC9M6SP02)					
Recognise and use combinations of transformations to create tessellations and other geometric patterns, using dynamic geometric software where appropriate (AC9M6SP03)					
Statistics and Probability					
Interpret and compare data sets for ordinal and nominal categorical, discrete and continuous numerical variables using comparative displays or visualisations and digital tools; compare distributions in terms of mode, range and shape (AC9M6ST01)					
Identify statistically informed arguments presented in traditional and digital media; discuss and critique methods, data representations and conclusions (AC9M6ST02)					
Plan and conduct statistical investigations by posing and refining questions or identifying a problem and collecting relevant data; analyse and interpret the data and communicate findings within the context of the investigation (AC9M6ST03)					
Recognise that probabilities lie on numerical scales of 0–1 or 0%–100% and use estimation to assign probabilities that events occur in a given context, using common fractions, percentages and decimals (AC9M6P01)					
Conduct repeated chance experiments and run simulations with an increasing number of trials using digital tools; compare observations with expected results and discuss the effect on variation of increasing the number of trials (AC9M6P02)					

Units

6 7 8 9 10 11 12 13 14 15 16 17 18 19 20 21 22 23 24 25 26 27 28 29 30 31 32 33 34 35

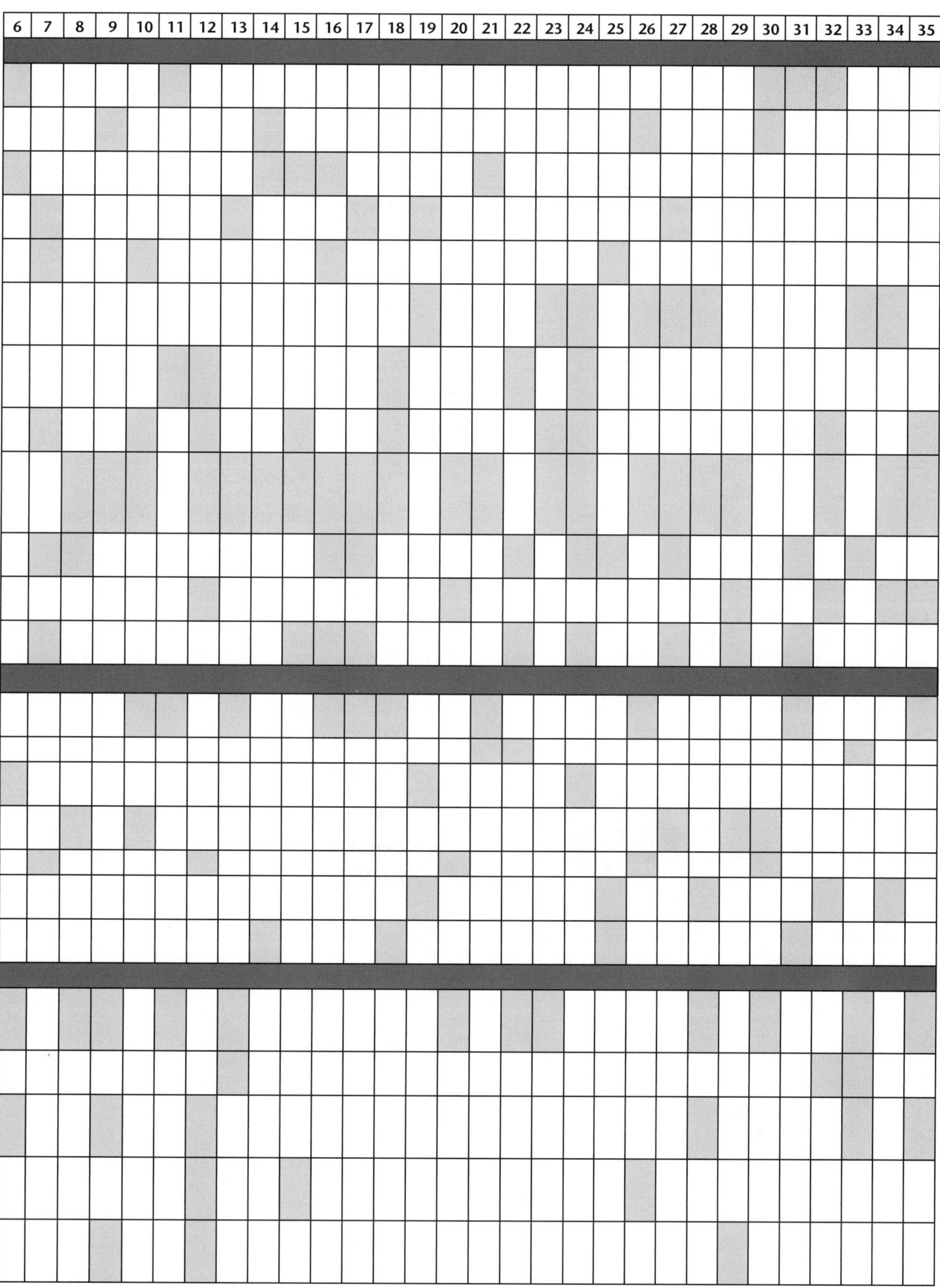

UNIT
1

Number and Algebra

SET 1 Basic

1 24 + 8

2 8×6

3 34 – 17

4 8 ☐ 4 = 32

5 27 ☐ 9 = 3

6 9^2

7 Divide 72 by 8.

8 12 ☐ 4 = 48

9 Product of 11 and 5

10 Sum of 9, 5 and 10

11 Quotient of 54 and 6

12 Cents in $8.69

13 $(6 + 2) \times (3 + 6)$

14 Factors of 45

15

SET 2 Place value

Draw beads on the abacuses to represent the numbers.

1 257 379

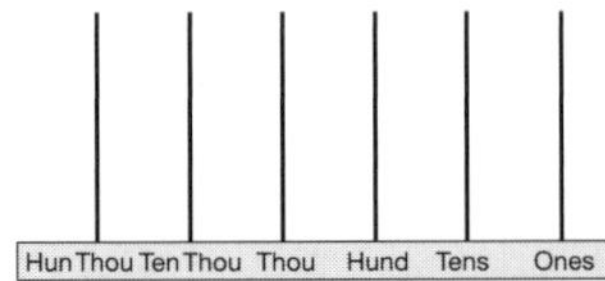

2

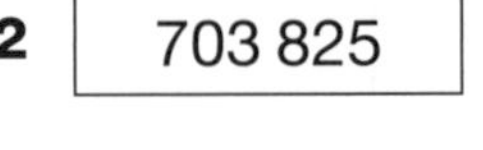

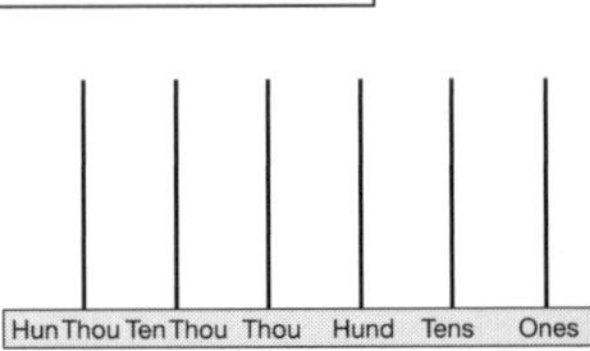

3 Write the number sixty-four thousand, nine hundred and twenty-eight.

4 What are the next two numbers after 52 817?

5 How many whole dollars in 19 361 cents?

6 Round 46 985 to the nearest thousand.

7 Subtract 1000 from 158 695.

8 What number is ten thousand less than 262 340?

9 How many hundreds are in 34 972?

10 What number is one thousand more than 442 186?

11 What are the two numbers before 214 901?

12 What is the next odd number after 25 131?

13 What number is ten thousand less than 985 970?

Space Angles

Name each angle.

1 ________

2

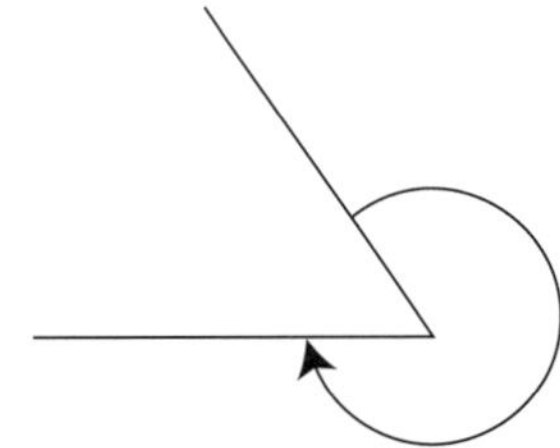

____

3

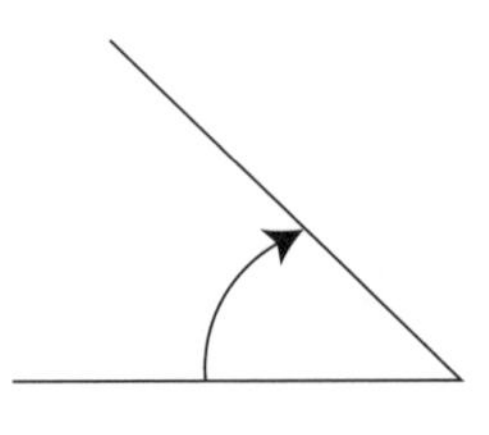

____

4

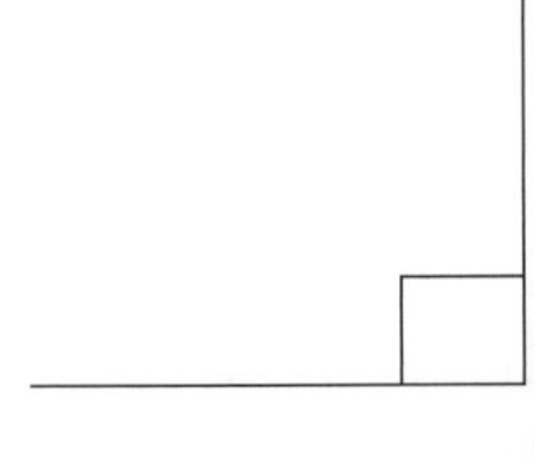

____

Number and Algebra

SET 3 Addition and subtraction strategies

Add these numbers mentally.

1 37 + 49 =

2 68 + 75 =

3 81 + 67 =

4 99 + 36 =

5 121 + 56 =

6 379 + 85 =

7 741 + 126 =

8 899 + 257 =

Subtract these numbers mentally.

9 76 – 29 =

10 89 – 57 =

11 68 – 34 =

12 149 – 37 =

13 259 – 58 =

14 741 – 127 =

Mathematical Reasoning

Round each number to the nearest 100 to estimate an approximate answer.

	Question	Rounded to 100	Approximate answer
15	395 + 206	400 + 200	600
16	591 – 298		
17	513 + 387		
18	785 – 589		
19	372 + 329		
20	882 – 286		

SET 4 Extension

1 5.8 + 3.6

2 20 × 5 × 8

3 Value of 8 in 78 321

4 $\frac{4}{10}$ of 50

5 Round 8.3 to the nearest whole number.

6 Add 1.7 to 4.4.

7 Five tickets at $1.65 each

8 Subtract the sum of 50 and 40 from 200.

9 Add the product of 6 and 8 to 42.

10 Round 1561 to the nearest thousand.

11 $\frac{7}{100}$ is less than 0.5 True or false?

12 If the water in the jug is 87°C, how many more degrees before it boils?

13 If Peter ran 0.8 km each day, how far did he run in a school week?

14 Round and estimate: 399 × 19.

15 If I spent $2, $1 and 65c, how much change would I receive from $5?

16 How many 250 g bags of rice can be made from a $3\frac{3}{4}$ kg bag?

Statistics and Probability Picture graphs

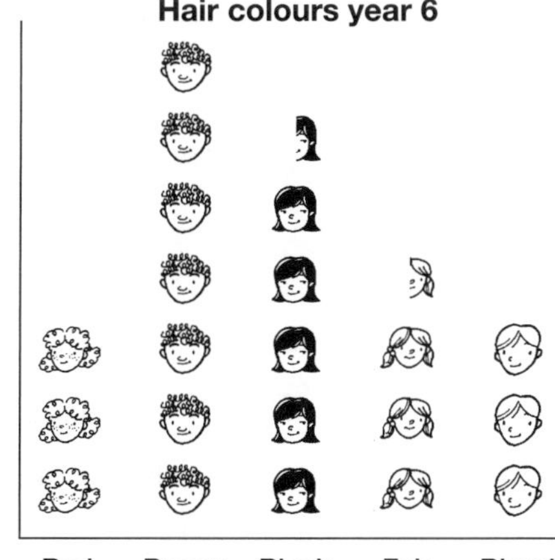

KEY 1 face = 4 children

Use the key to answer the questions.
How many children had:

1 red hair? ____________

2 brown hair? ____________

3 black hair? ____________

4 fair hair? ____________

5 blond hair? ____________

Number and Algebra

SET 1 Basic

1 ☐ – 8 = 3

2 85 × 0

3 21 ÷ 3

4 400 – 150

5 7^2

6 Date after 11 February

7 49 ÷ 7

8 5c × 100

9 Difference between 19 and 90

10 Season before spring

11 28 ☐ 7 = 4

12 Quotient of 24 and 6

13 How many quarters in 5?

14 Value of 7 in 17 940

15

What is 25 cm less than a metre?
☐ cm

SET 2 Square numbers

Calculate the area of the squares.

1 3 cm

A = ____cm²

2 4 cm

A = ____cm²

3 5 cm

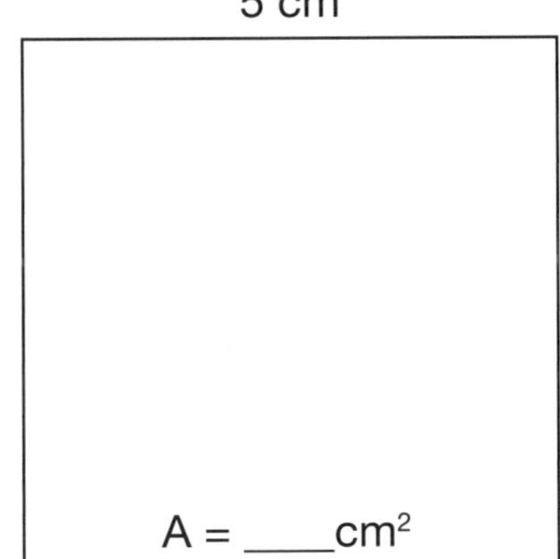

4 2 cm

A = __cm²

5 6^2 =

6 7^2 =

7 8^2 =

8 10^2 =

9 1^2 =

10 9^2 =

Space Polygons

Colour code the shapes to their matching labels.

All my angles are obtuse. I have 5 axes of symmetry.

I have 6 right angles and 2 reflex angles. I have 8 sides.

I have 0 right angles and 0 lines of symmetry. I have 10 sides.

I have 6 sides and 6 angles all the same size. I have 6 axes of symmetry.

Number and Algebra

SET 3 Negative numbers

Record the temperature shown in each thermometer.

1 °C 55° 50° 45° 40° 35° 30° 25° 20° 15° 10° 5° 0° –5° –10° –15° –20° –25°

☐ °C

2 °C 55° 50° 45° 40° 35° 30° 25° 20° 15° 10° 5° 0° –5° –10° –15° –20° –25°

☐ °C

3 °C 55° 50° 45° 40° 35° 30° 25° 20° 15° 10° 5° 0° –5° –10° –15° –20° –25°

☐ °C

Use the number line to solve the problems.

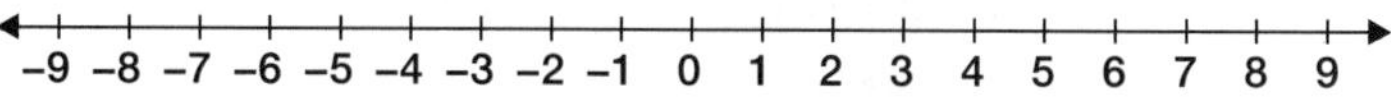

4 $3 + 5 =$

5 $3 - 5 =$

6 $-3 + 5 =$

7 $-4 + 7 =$

8 $2 - 6 =$

9 $4 - 8 =$

10 $2 - 8 + 6 =$

11 $1 - 5 + 7 =$

SET 4 Extension

1 6×15

2 $\frac{23}{100} = \square\%$

3 Value of 6 in 37 206

4 Average of 45, 25, 35

5 $\frac{3}{4}$ of $96

6 How many faces has a rectangular prism?

7 Round 63 209 to the nearest 100.

8 If 3 kg cost $18.60, how much would 5 kg cost?

9 $\frac{38}{100} = 38\%$ True or false?

10 $300 \div 2 \times 10 - 750 + 53$

Mathematical Reasoning

Solve these missing number sentences. Each shape has the same value in every question. For example, the square is always equal to 9.

11 $\square \times \triangle = 54$

12 $⬡ \times 2 = \triangle$

13 $\square \div ⬡ = ⬡$

14 $\square \times ⬡ + \triangle = 33$

15 $72 \div \square \times \triangle = 48$

16 $54 \div \triangle = ⬡ \times ⬡$

Measurement Perimeter

Measure the edges of the shapes in millimetres. Record the perimeter in the boxes.

1 ☐ mm

2 ☐ mm

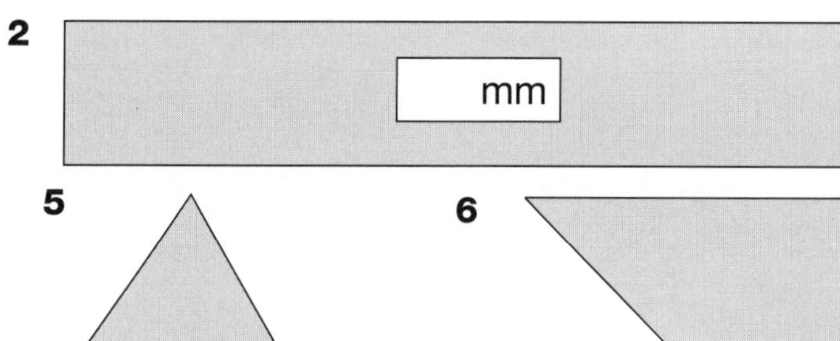

3 ☐ mm

4 ☐ mm

5 ☐ mm

6 ☐ mm

Number and Algebra

SET 1 Basic

1 8 + 7

2 3 × 4

3 13 – 4

4 6 ☐ 3 = 18

5 18 ☐ 6 = 12

6 3^2

7 Divide 12 by 2.

8 18 ☐ 3 = 6

9 Product of 7 and 4

10 Sum of 7 and 14

11 Difference between 14 and 21

12 (3 + 7) × 7

13 Is 28 a multiple of 7?

14 1207, 1210, 1213, ☐

15

Peter has run 550 m of an 800 m race. How much further does he need to run to finish the race?

☐ m

SET 2 Factors

Answer true or false.

1 3 is a factor of 12.

2 7 is a factor of 21.

3 8 is a factor of 30.

4 4 is a factor of 20.

5 4 is a factor of 25.

6 6 is a factor of 36.

7 9 is a factor of 81.

8 7 and 3 are both factors of 21.

9 9 and 4 are both factors of 36.

10 5 and 6 are both factors of 35.

11 8 and 4 are both factors of 48.

12 10 and 5 are both factors of 100.

13 3 and 9 are both factors of 36.

14 7 and 4 are both factors of 56.

15 Write a rule for 5 being a factor of a number.

____

____

Space Classifying three-dimensional objects

Complete the grid to classify the shapes.

1

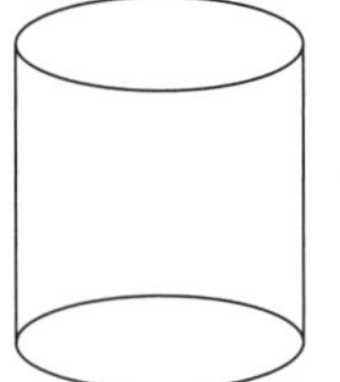

2

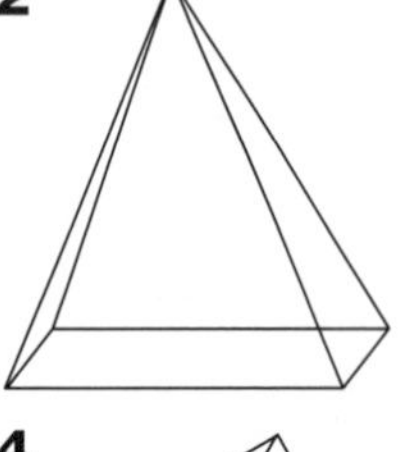

3

4

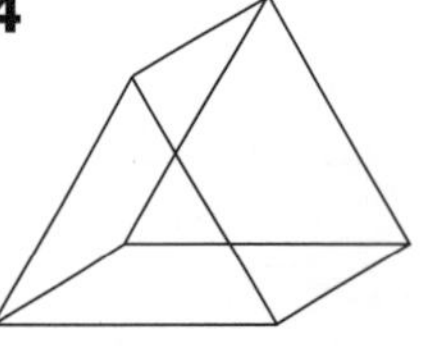

	Name	Faces	Vertices	Edges
1				
2				
3				
4				

Number and Algebra

SET 3 Percentages, fractions and decimals

	Fraction	Decimal	%
1	$\frac{10}{100}$		
2	$\frac{25}{100}$		
3	$\frac{7}{10}$		
4	$\frac{20}{100}$		
5	$\frac{1}{2}$		
6	$\frac{1}{4}$		
7	$\frac{3}{4}$		

Order from smallest to largest.

8	$\frac{27}{100}$	30%	0.29	
9	35%	$\frac{53}{100}$	0.33	
10	$\frac{99}{100}$	9%	0.9	
11	0.54	$\frac{1}{2}$	49%	
12	4%	$\frac{3}{10}$	0.21	
13	$\frac{9}{100}$	90%	0.95	
14	$\frac{70}{100}$	7%	0.03	

Mathematical Reasoning

SET 4 Extension

1 Value of 7 in 73 256

2 Write $\frac{1}{100}$ as a percentage.

3 1.8 m = ☐ cm

4 Round 3.9 to the nearest whole number.

5 Average 43, 36, 50

6 3113, 3128, 3143, ☐

7 (20 + 18) × 2

8 Would a hot summer's day be 20°C, 35°C or 100°C?

9 How many grams in 7.3 kg?

10 Round 29 306 to the nearest 1000.

11 $\frac{3}{10}$ of 200

12 Perimeter of a pentagon with sides of 1.5 cm

13 How many axes of symmetry has a square?

14 $\frac{3}{5} = \frac{\square}{40}$

15 Tom ate 25% of the cake.
Jill ate 0.27 of it and
Max ate 33 hundredths of it.
How much was left? 

16 How much does 9 kg of meat cost at $3.50 kg? 

Measurement Square centimetres

Calculate the area of these rectangles in square centimetres (cm^2).

1

2

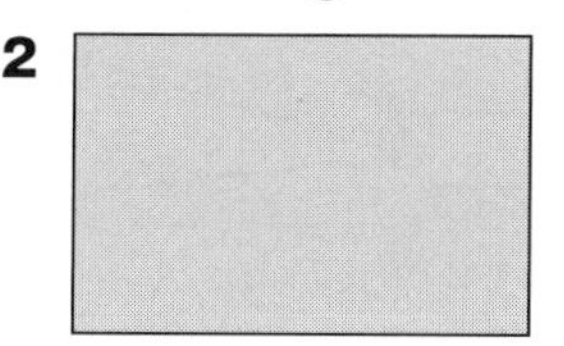

3

	Length	× Width	= Area
1			
2			
3			

Number and Algebra

SET 1 Basic

1 9 × 5

2 16 – 9

3 14 + 8

4 4^2

5 16 ☐ 4 = 12

6 18 ☐ 3 = 6

7 Divide 35 by 7.

8 20 ☐ 20 = 40

9 $\frac{1}{4}$ × \$28

10 Quotient of 25 and 5

11 Factors of 9

12 Are 18 and 24 multiples of 6?

13 Is 21 a prime number?

14 How many cm in 7 m?

15 Phillip spent \$6.50 and \$4.50. What was the total amount he spent?

\$ ☐

SET 2 Division

1 $3\overline{)969}$

2 $4\overline{)568}$

3 $5\overline{)175}$

4 $6\overline{)156}$

5 $7\overline{)763}$

6 $8\overline{)744}$

7 Share \$64 among 8

8 How many fives in 90?

9 450 ÷ 10

10 What is the quotient when 357 is divided by 7?

11 Share 424 blocks between 4 groups.

12 Lisa's mum won \$924 in the lottery. If she shared it with another 6 people, how much did each person receive? 

13 540 spectators were seated in 10 rows. How many in each row? 

14 A group of 217 boys was divided into 7 teams. How many in each team? 

Probability and Statistics Tree diagram

Complete the tree diagram to show all the possible outcomes that could occur if a coin is tossed three times.

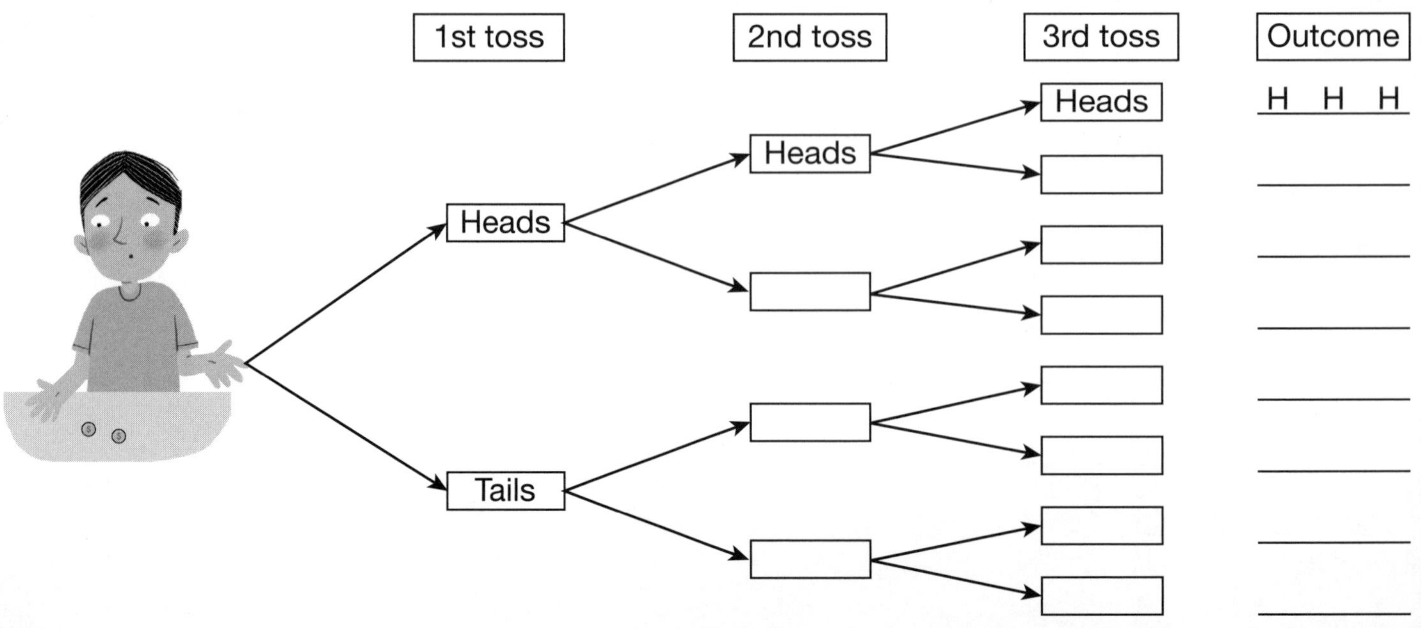

Number and Algebra

SET 3 Number patterns

Follow the rule to complete the number patterns.

1 Rule: Multiply by 6

2	4	6	8	10	12	14	16	18
12								

2 Rule: Subtract 12

120	108	96	84	72	60	48	36	24

3 Rule: Divide by 8

96	88	80	72	64	56	48	40	32
12								

4 Rule: Add 5^2

10	20	30	40	50	60	70	80	90

5 Make your own rule and supply your own numbers.

Rule:

SET 4 Extension

1 160 – 85

2 $\frac{3}{4} = \frac{\square}{12}$

3 Value of 7 in 86 791

4 Complete this sequence: 5404, 5504, 5604, ☐, ☐

5 How many months in half a year?

6 $\frac{3}{4} \times 48 + 207$

7 Round $18.61 to the nearest dollar.

8 Average of 45, 26, 39 and 450

9 180 seconds = ☐ minutes

10 How much is 4 kg at $1.70 a kg?

11 1 km ÷ 4 = ☐ m

12 Place in a sequence: $\frac{1}{2}$, $\frac{1}{4}$, $\frac{1}{3}$, $\frac{3}{4}$

13 9:25 + 58 minutes

14 $0.72 × 9

15 How much change would I receive from $20 if I bought three $1.85 ice creams?

16 If 3 kg costs $1.95, how much would 10 kg cost?

17 Round off to estimate an answer to 51 × 494.

Measurement The cubic centimetre

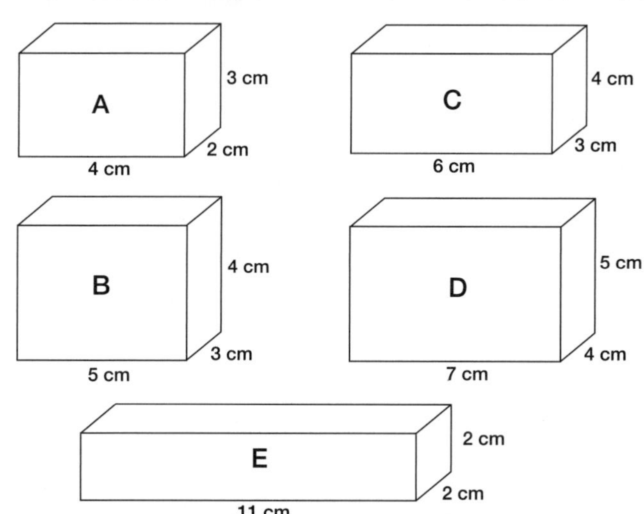

Calculate the volume of these prisms.

	Shape	Length	Width	Height	Volume
1	A				
2	B				
3	C				
4	D				
5	E				

Number and Algebra

SET 1 Basic

1 18 + 6

2 4×9

3 25 – 16

4 5 ☐ 8 = 40

5 9 ☐ 3 = 3

6 8^2

7 Divide 64 by 8.

8 11 ☐ 3 = 33

9 Product of 10 and 6

10 Sum of 8, 6 and 9

11 Quotient of 54 and 9

12 Cents in $9.61

13 $(4 + 5) \times (4 + 6)$

14 Factors of 27

15

What are the next two numbers in this sequence?
27, 54, 108,

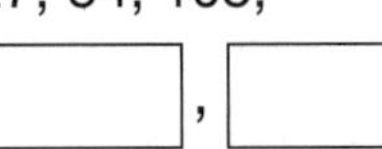

SET 2 Multiplication strategies

Calculate mentally.

1 9×3

2 3×90

3 3×900

4 30×90

5 70×9

6 7×800

7 30×50

8 60×70

Estimate the answers by rounding the larger numbers.

9 29×3

10 49×4

11 57×6

12 $99 \div 5$

13 71×5

14 39×20

15 52×30

16 $199 \div 4$

Calculate mentally by multiplying the tens and the ones separately, then combining them.

17 35×4

18 42×5

19 55×6

20 44×4

21 27×8

Space Rotational symmetry

Colour the shapes that have rotational symmetry.

1

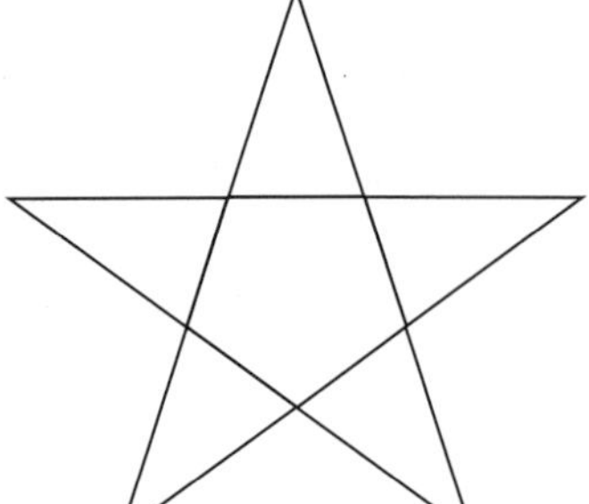

2

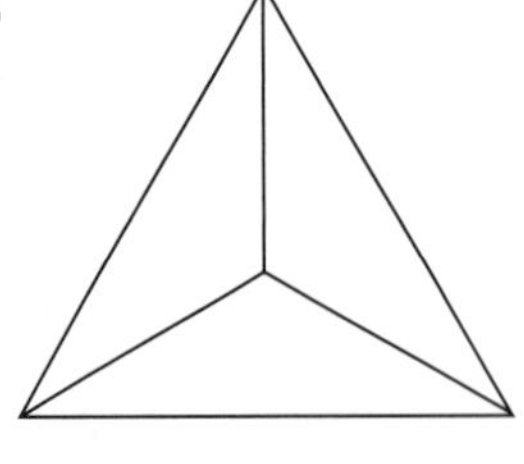

3

4

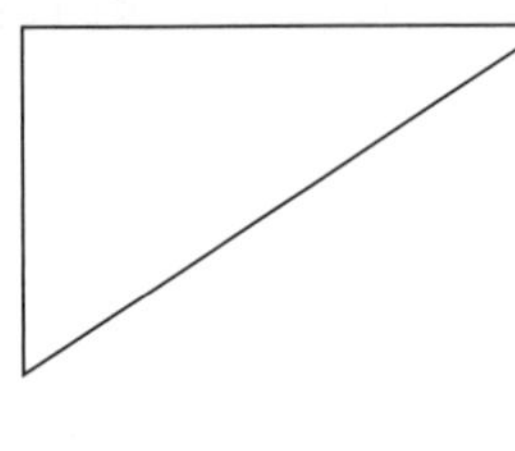

5

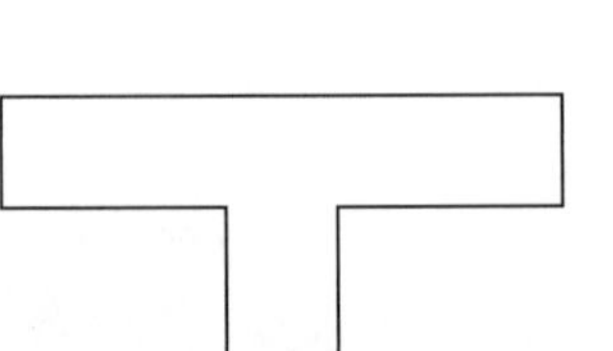

6

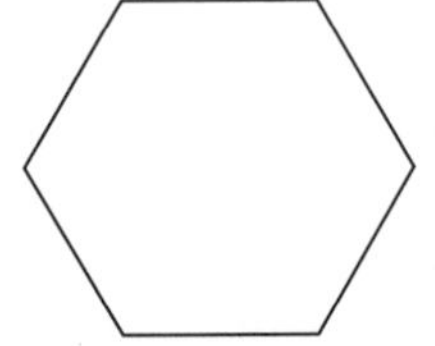

Number and Algebra

SET 3 Subtracting decimals

1 0.6 – 0.3

2 0.9 – 0.4

3 0.89 – 0.74

4 0.77 – 0.53

5 0.86 – 0.34

6 0.99 – 0.55

7 $\begin{array}{r} 8.74 \\ -\ 1.32 \\ \hline \end{array}$

8 $\begin{array}{r} 7.43 \\ -\ 1.31 \\ \hline \end{array}$

9 $\begin{array}{r} 36.46 \\ -\ 13.51 \\ \hline \end{array}$

10 $\begin{array}{r} 74.94 \\ -\ 11.38 \\ \hline \end{array}$

11 $\begin{array}{r} 423.5 \\ -\ 116.3 \\ \hline \end{array}$

12 $\begin{array}{r} 294.38 \\ -\ 36.19 \\ \hline \end{array}$

SET 4 Extension

1 440 – 92

2 $\frac{3}{4} + \frac{3}{4} + \frac{1}{4}$

3 Value of 6 in 96 401

4 Complete this sequence:
6924, 6914, 6904, ☐, ☐

5 At what temperature does water begin to boil?

6 How many dollars in 19 615 cents?

7 Order $\frac{3}{5}$, $\frac{1}{2}$, 0.47 and 40%.

8 6 hours = ☐ minutes

9 How much change would I receive from $50 if I spent $3.28?

10 How far did we travel if we drove for $4\frac{1}{2}$ hours at an average speed of 68 km/h?

11 What is the third angle of a triangle if the other two are 27° and 58°?

12 How many m^2 in 1 hectare?

13 5 × 9 + 2 × 9 + 3 × 9

14 $\frac{3}{5}$ of 200 ÷ $\frac{3}{4}$ of 16

15 How much are 13 brushes at $3.75 each?

16 What is the perimeter of an octagon with 115 mm sides?

Statistics and Probability Probability from 0 to 1

1 Describe the chance of pulling each coin out of the bag by drawing a line to its probability on the chance scale.

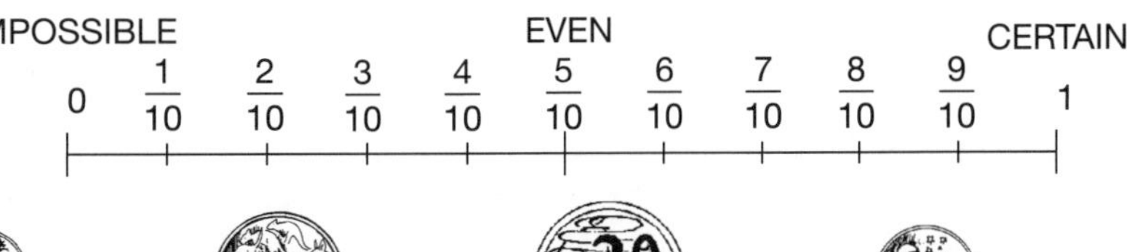

2 Describe the chance of pulling 2 coins out of the bag that have a combined value of less than 50c.

Number and Algebra

SET 1 Basic

1 100 ☐ 25 = 75

2 8 × 3

3 13 – 6

4 24 ÷ 4

5 80 + 80

6 100 ☐ 120 = 220

7 100 ☐ 2 = 50

8 Product of 7 and 9

9 Difference between 24 and 6

10 Value of 6 in 6135

11 Factors of 20

12 1405, 1411, 1417, ☐

13 Is 9 a prime number?

14 $3\frac{1}{2}$ min = ☐ seconds

15 How many 2 m lengths can be cut from 20 m?

 lengths

SET 2 Negative numbers

Use the thermometer to solve the problems.

1 10° C + 25° C =

2 10° C – 15° C =

3 –10° C + 15° C =

4 –10° C + 25° C =

5 –25° C + 30° C =

6 –15° C + 25° C =

7 10° C – 25° C =

8 5° C – 20° C =

9 15° C – 25° C =

10 – 5° C – 25° C =

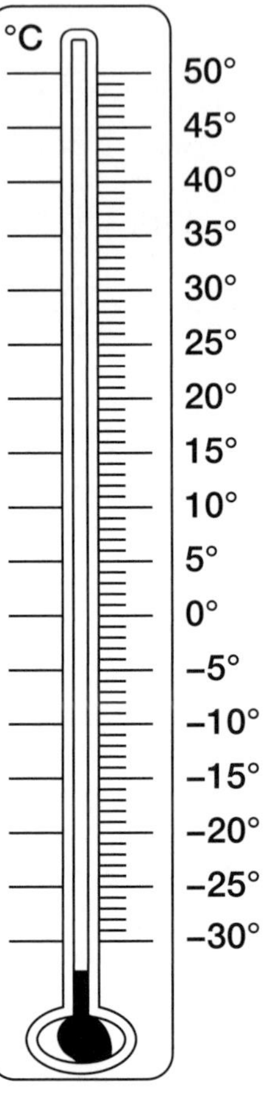

Statistics and Probability Column graphs

40 children who own pets were asked to name their favourite pet.

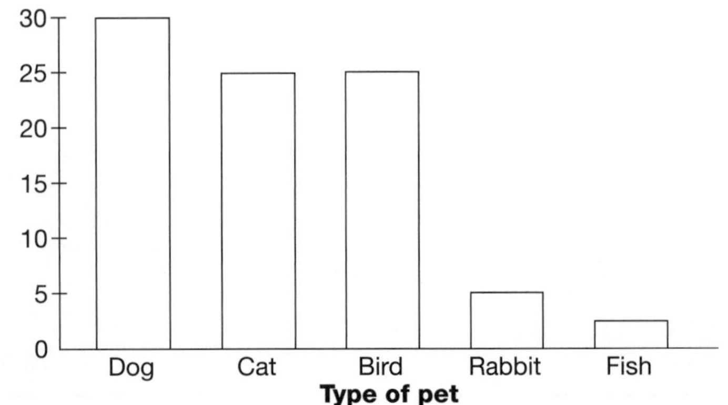

1 Which pet is the least popular?

____

2 How many named birds as their favourite?

____

3 Which pet represents the median?

____

4 Which pets are equally popular?

____

Number and Algebra

SET 3 Improper fractions and mixed numbers

True or false?

1 $\frac{3}{2} = 1\frac{1}{2}$ ☐

2 $\frac{5}{4} = 1\frac{1}{4}$ ☐

3 $\frac{15}{10} = 1\frac{5}{10}$ ☐

4 $\frac{6}{5} = 1\frac{1}{5}$ ☐

5 $\frac{4}{3} = 1\frac{2}{3}$ ☐

6 $\frac{9}{4} = 2\frac{1}{4}$ ☐

7 $\frac{12}{10} = 2\frac{2}{5}$ ☐

8 $\frac{8}{5} = 1\frac{4}{5}$ ☐

9 $\frac{7}{3} = 2\frac{1}{3}$ ☐

10 $\frac{7}{2} = 3\frac{1}{2}$ ☐

Convert each improper fraction to a mixed number.

11 $\frac{4}{3}$ =

12 $\frac{7}{5}$ =

13 $\frac{9}{4}$ =

14 $\frac{8}{6}$ =

15 $\frac{11}{5}$ =

16 $\frac{5}{2}$ =

Mathematical Reasoning

Name two improper fractions that are equal to each mixed number.

	Mixed number	Improper fraction	
17	$2\frac{1}{2}$		
18	$1\frac{1}{4}$		
19	$1\frac{1}{3}$		
20	$2\frac{1}{6}$		

SET 4 Extension

1 7 × 15

2 If 3 kg costs $3.54, how much would 6 kg cost?

3 Value of 4 in 1 400 962

4 Average of 41, 52, 18

5 Order 35%, 0.2, $\frac{1}{4}$ and 0.31.

6 How much is 3.5 kg at $8 per kg?

7 Round $19.25 to the nearest dollar.

8 2 km ÷ 4

9 (16 – 12) × 5

10 280 + 40 × 8

11 How much change would I receive from $50 if I spent $7.18?

12 $\frac{7}{10}$ of a kilometre

13 $\frac{1}{10} + \frac{17}{100} + \frac{9}{100} =$

14 $\frac{3}{8} + \frac{7}{8} + \frac{5}{8} =$

15 Round 9.32 to the nearest tenth.

16 $\frac{3}{5} \times 75 + \frac{2}{5} \times 50$

Mathematical Reasoning

Complete the decimal addition chart.

	+	7.431	5.323	2.249
17	0.05			
18	0.003			
19	0.5			
20	0.001			

Measurement Timetables

1 What time does the 8:30 am bus from Barton Street arrive at Belgrave Street?

2 Where is the 7:55 bus from Barton Street 27 minutes later?

3 How long until the next bus after the 7:35 from Kyle Station?

4 How long does it take the 11:31 bus from Jubilee Oval to reach Belgrave Street?

5 What bus from Barton Street would I need to catch to arrive at Austral Street before 11 am?

Barton Street	Jubilee Oval	Kyle Station	Byrnes Street	Stuart Street	Belgrave Street	Austral Street
						AM SERVICE
–	–	–	–	–	–	6:30
–	–	–	–	–	6:55	7:00
–	7:00	7:05	7:15	7:22	7:25	7:30
–	7:30	7:35	7:45	7:52	7:55	8:00
7:55	7:56	8:05	8:15	8:22	8:25	8:30
8:30	8:31	8:35	8:45	8:52	8:55	9:00
–	9:00	9:05	9:15	9:22	9:25	9:30
–	9:30	9:35	9:45	9:52	9:55	10:00
10:00	10:01	10:05	10:15	10:22	10:25	10:30
–	10:30	10:35	10:45	10:52	10:55	11:00
–	11:00	11:05	11:15	11:22	11:25	11:30
11:30	11:31	11:35	11:45	11:52	11:55	12:00

UNIT
7

Number and Algebra

SET 1 Basic

1 $48 \div 6$

2 $(4 + 3) \times 5$

3 $4 + 4 + 4 + 4$

4 7^2

5 Product of 9 and 2

6 Cents in \$4.60

7 $(4 + 3) \times (5 + 1)$

8 3c × 100

9 56 ☐ 8 = 7

10 Divide 90 by 10.

11 Season after summer

12 Is 29 a multiple of 7?

13 Quotient of 63 and 7

14 $(6 + 2) \times 4$

15

How do you write one hundred and ninety-two in Hindu-Arabic? ☐

SET 2 Add and subtract fractions

Add these fractions.

1 $\frac{3}{8} + \frac{2}{8} =$

2 $\frac{7}{10} + \frac{1}{10} =$

3 $\frac{3}{10} + \frac{3}{10} + \frac{3}{10} =$

4 $\frac{3}{8} + \frac{3}{8} + \frac{1}{8} =$

5 $\frac{7}{8} + \frac{5}{8} = \quad =$

6 $\frac{3}{4} + \frac{3}{4} = \quad =$

7 $\frac{7}{10} + \frac{7}{10} = \quad =$

8 $\frac{3}{5} + \frac{4}{5} = \quad =$

9 $\frac{5}{10} + \frac{5}{10} + \frac{5}{10} = \quad =$

10 $\frac{5}{12} + \frac{4}{12} + \frac{4}{12} = \quad =$

Subtract these fractions.

11 $\frac{7}{10} - \frac{1}{10} =$

12 $\frac{7}{8} - \frac{2}{8} =$

13 $\frac{9}{10} - \frac{5}{10} =$

14 $\frac{4}{5} - \frac{3}{5} =$

15 $\frac{3}{4} - \frac{2}{4} =$

16 $\frac{5}{10} - \frac{2}{10} =$

17 $\frac{4}{8} - \frac{3}{8} =$

18 $\frac{6}{10} - \frac{1}{10} =$

19 $\frac{2}{5} - \frac{1}{5} =$

20 $\frac{9}{10} - \frac{3}{10} =$

Space Representing three-dimensional objects

Draw these prisms on the dot paper. The back block has been drawn for you.

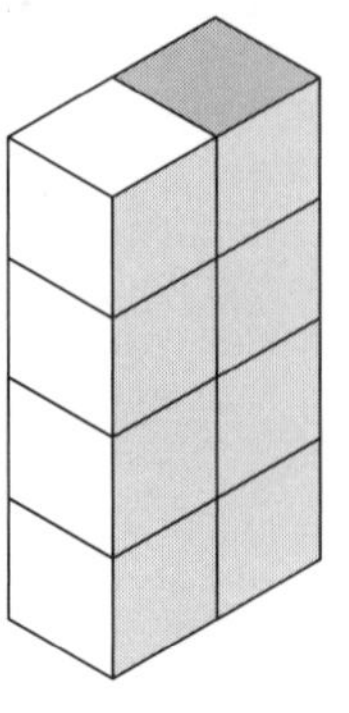

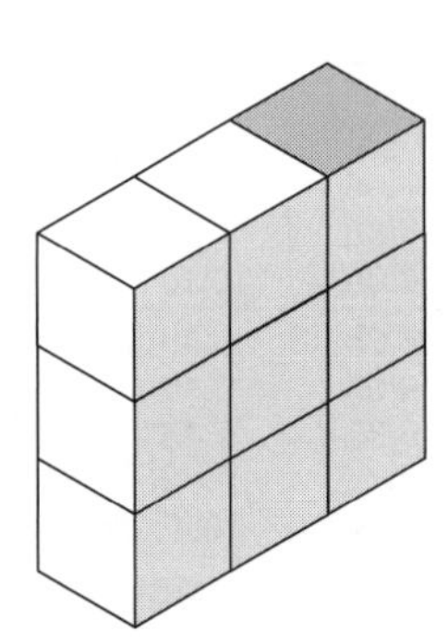

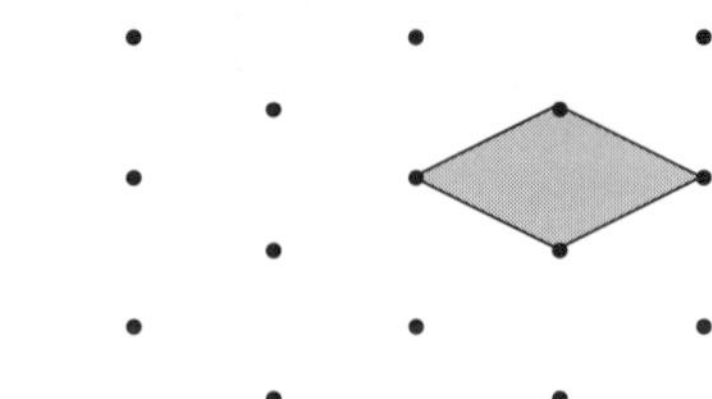

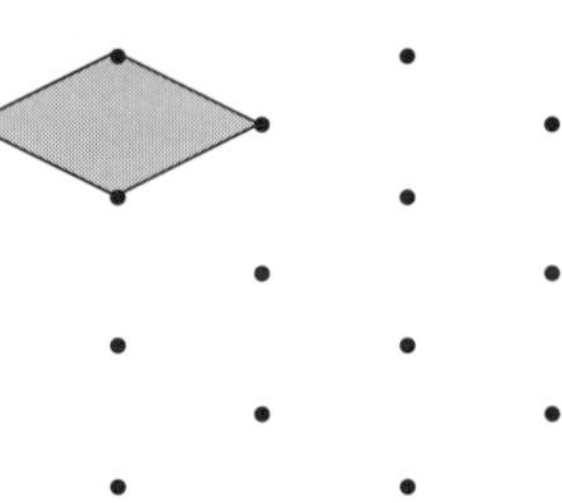

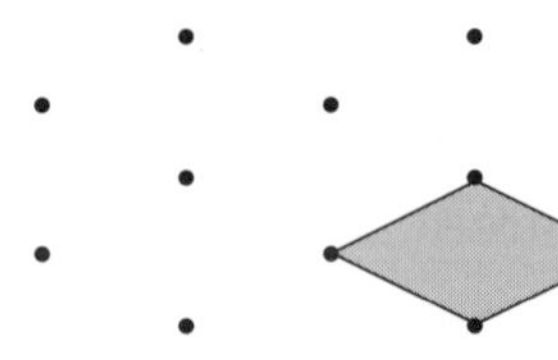

Number and Algebra

SET 3 Decimal place value

State the place value of each bold digit.
The words in the cloud may help you.

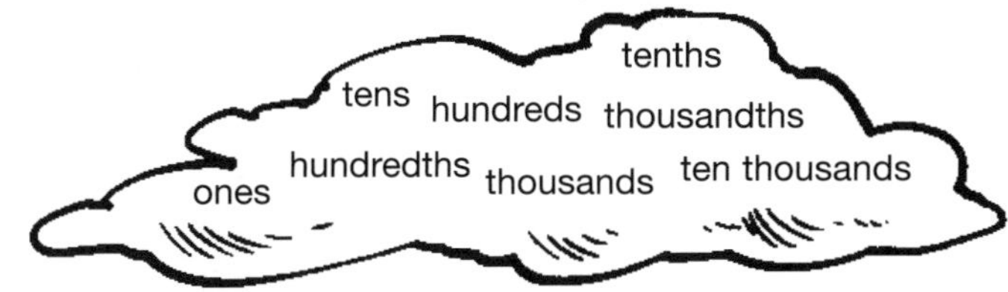

	Number	Place value
1	36**7**.361	
2	3**6**5.973	
3	**7**93.548	
4	357.**2**42	
5	679.3**8**4	
6	**3**574.261	
7	4729.3**7**5	
8	293.74**2**	

Convert the measurements into smaller units.

9 7.521 kg = __________ grams

10 5.425 L = __________ millilitres

11 1.785 m = __________ millimetres

12 6.301 t = __________ kilograms

SET 4 Extension

1 1.25 km = ☐ m

2 Share $357 between six.

3 7×501

4 How many eggs in $9\frac{1}{2}$ dozen?

5 How many faces does a triangular prism have?

6 What is the perimeter of a hexagon with 16 cm sides?

7 Centimetres in 13.75 m

8 Which one is not equivalent: $\frac{1}{4}$, 25%, 0.35 or $\frac{25}{100}$?

Mathematical Reasoning

Round these numbers to the nearest 100 to estimate an answer to the multiplications.

	Number	Rounded to	Multiplied by	Estimate
9	321	300	4	1200
10	494		8	
11	340		6	
12	897		3	
13	515		5	
14	899		9	

Number and Algebra Decimal number patterns

Complete the function machines.

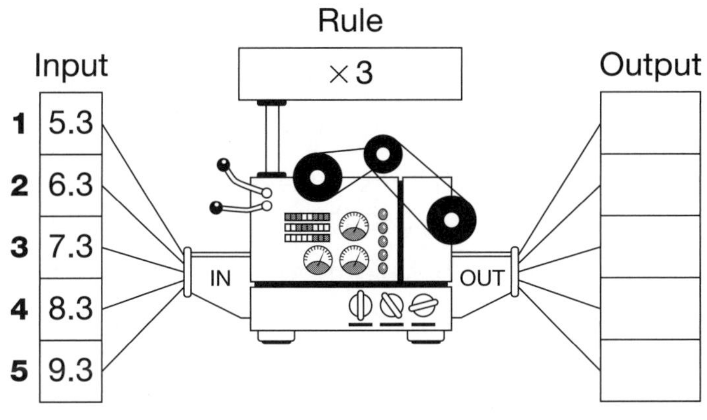

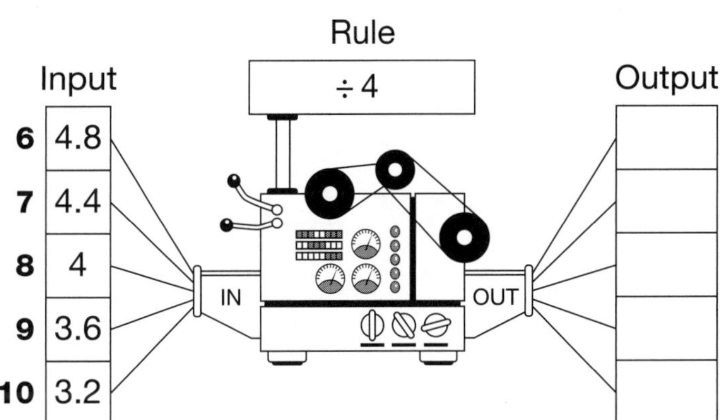

UNIT 8

Number and Algebra

SET 1 Basic

1 Divide 40 by 5.

2 Product of 7 and 5

3 17 + 7

4 3 × 9

5 19 ☐ 8 = 27

6 27 ☐ 3 = 9

7 38 – 8

8 21 ÷ 7

9 9 ☐ 3 = 27

10 4 m = ☐ cm

11 Which is greater, $\frac{7}{10}$ or $\frac{7}{100}$?

12 Factors of 15

13 Are 35 and 50 multiples of 5?

14 How many 10c lollies can I buy for $1.50?

15

SET 2 Problem solving/large numbers

1
```
   43 564
   57 277
 + 49 802
```

2
```
   76 805
   75 957
 + 21 965
```

3
```
   75 384
 – 24 275
```

4
```
   76 854
 – 46 928
```

5
```
  $75 862
 – 29 427
```

6
```
  $94 386
 – 38 279
```

7 Mrs Collin's new job pays a salary of $78 565 as well as a car allowance of $12 999. What is the value of her yearly wage?

8 When Mr O'Neill prepared his tax return he calculated that during the year he earned $67 324 but he would have to pay $28 949 in tax. How much did he earn after tax?

Statistics and Probability Line graphs

The graph below shows the weight of an average boy at ages 0 to 16.

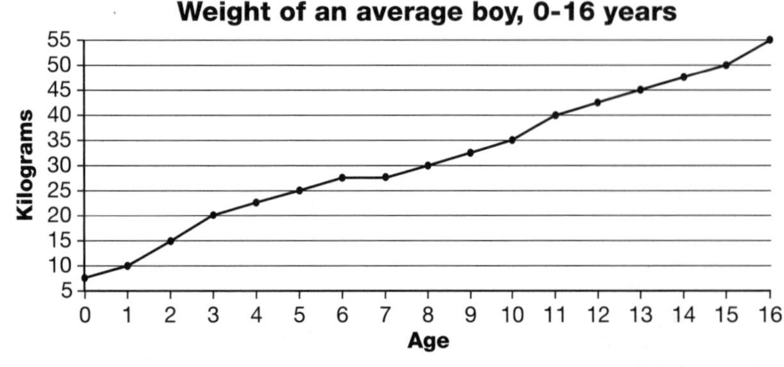

1 How much does an average boy weigh at age 5?

2 How much does an average boy weigh at age 13?

3 Between which ages does an average boy's weight stay the same?

4 How much weight does an average boy put on between the ages of 5 and 11?

5 How much weight does an average boy put on between the ages of 10 and 16?

6 Estimate an average boy's weight at $10\frac{1}{2}$ years.

Number and Algebra

SET 3 Triangular numbers

Use the dot paper to find out if the following are triangular numbers.

1 6 ☑	2 15 ☐	3 21 ☐

4 30 ☐	5 36 ☐

SET 4 Extension

1 Write 13.25 as a mixed number.

2 Average 46, 23, 69, 22

3 How many kg in 3.1 tonnes?

4 Which is larger, $\frac{3}{4}$ or $\frac{3}{10}$?

5 Which temperature would be a cool winter's day: 41°C, 30°C, 100°C or 15°C?

6 Write one million, two hundred and twenty-six thousand in figures.

7 Share $8448 among 4 people.

8 How many faces has a triangular pyramid?

9 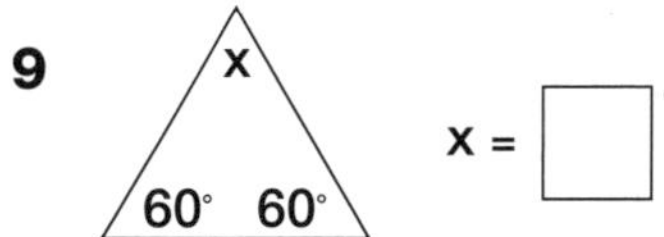

10 Order 3.5, $4\frac{1}{4}$, 3.75 and $3\frac{9}{10}$.

11 How much is 3.2 kg of meat at $10 per kg?

12 How many axes of symmetry has a hexagon?

Mathematical Reasoning

13 How many boys and girls are in our class? $\frac{2}{3}$ of the class are girls and half of the girls in our class make a squad of 11 netball players.

____

Space Constructing angles

Use a protractor to create these angles.

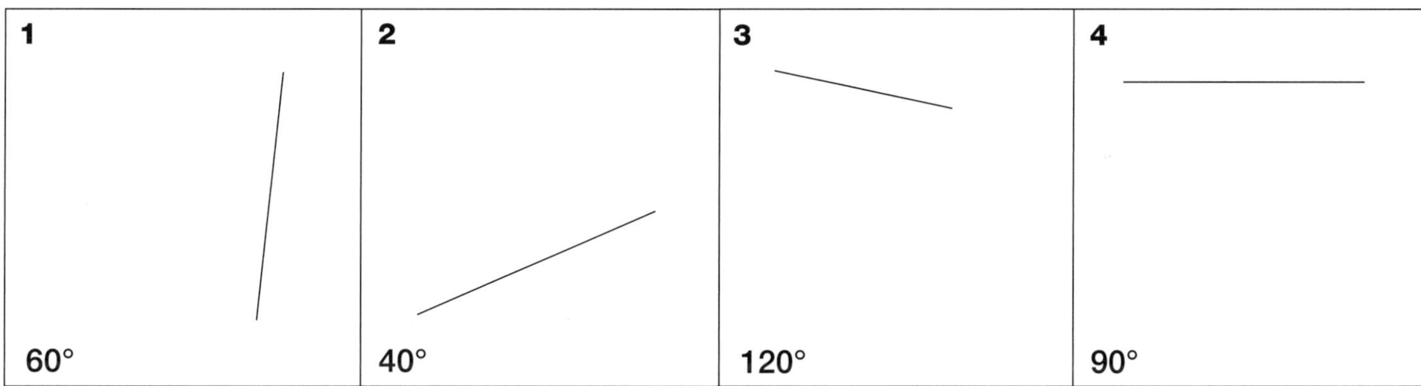

Number and Algebra

SET 1 Basic

1 7^2

2 3 ☐ 8 = 24

3 35 ☐ 5 = 7

4 8 × 8

5 36 – 8

6 37 + 8

7 42 ÷ 6

8 Product of 9 and 7

9 Sum of 15 and 23

10 Cents in $37.52

11 Are 21 and 16 multiples of 6?

12 $2\frac{3}{4}$ hours = ☐ minutes

13 Which is greater, 27 hundredths or 30%?

14 55c × 6

15

SET 2 Division

Solve these problems.

1 I paid $5250 for 5 televisions. How much were they each? 

2 A car travelled 1835 km in 5 days. What was the average distance travelled each day? 

3 6930 bags were packed into 10 boxes. How many bags in each box? 

4 John had a bag of 981 stamps to share between himself and 2 friends. How many stamps did each person receive? 

5 Melissa earned $2352 over 4 weeks. What was her average weekly earnings? 

6 1000 mL of perfume was divided into 8 mL sample bottles. If each bottle was sold for $5, what was their total value?  

Statistics and Probability Chance

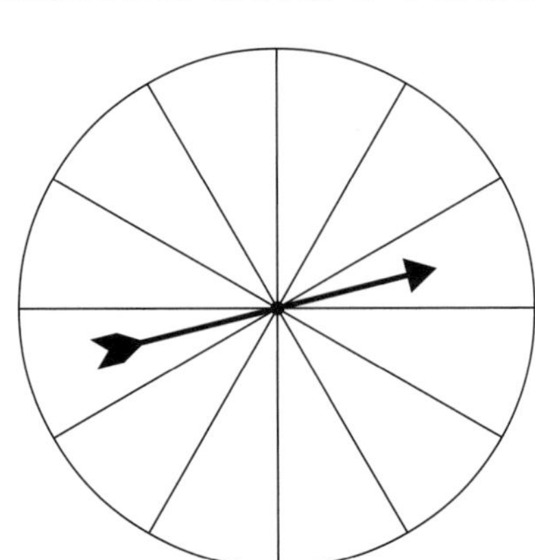

Colour the spinner to reflect the chances.

1 $\frac{1}{2}$ of the chances are green.

2 $\frac{1}{12}$ of the chances are yellow.

3 $\frac{1}{4}$ of the chance are red.

4 $\frac{2}{12}$ of the chances are blue.

Number and Algebra

SET 3 Prime and composite numbers

Write prime or composite after each number.

1 12 ______________________

2 7 ______________________

3 16 ______________________

4 21 ______________________

5 45 ______________________

6 49 ______________________

7 92 ______________________

8 95 ______________________

9 75 ______________________

10 50 ______________________

11 17 ______________________

12 57 ______________________

Write all the factors of these composite numbers.

13	18	
14	21	
15	32	
16	56	

SET 4 Extension

1 Centimetres in 3.7 m

2 30 923, 30 950, 30 977, ☐

3 How many faces has a pentagonal prism?

4 Average of 124, 148, 207, 21

5 Round 19 978 to the nearest 1000.

6 $1\frac{3}{4}$ min = ☐ seconds

7 33 + 27 ÷ 3 − 9 =

8 Which one is not equivalent: 0.5, 50%, $\frac{1}{2}$ or 0.51?

9 2.35 km = ☐ m

10 Write two hundred and ninety thousand and six in figures.

11 If six erasers cost $2.40, how much would five cost?

12 How much is 4.25 m of ribbon at $8 per metre?

Statistics and Probability Graphs

Calculate the differences in temperature.

1 Minimum temperature of Auckland and Paris.

2 Maximum temperature of Moscow and London.

3 Maximum and minimum temperatures of Rome.

4 Minimum London and maximum Moscow.

5 Minimum Auckland and minimum Moscow.

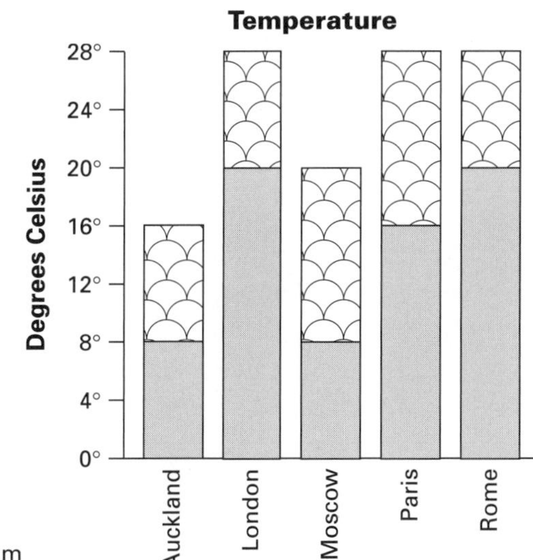

Number and Algebra

SET 1 Basic

1 7 + 9

2 8 × 7

3 24 – 9

4 49 ☐ 40 = 89

5 7 ☐ 10 = 70

6 5^2

7 Divide 35 by 5.

8 50 ☐ 25 = 25

9 Product of 7 and 6

10 Quotient of 45 and 5

11 Cents in $13.24

12 (8 + 1) × 3

13 Is 27 a prime number?

14 3 L and 175 mL = ☐ mL

15 How many 25 cm lengths can be cut from 1 m?

16

Dominic has 25 books to read in a Read-A-Thon. If he has read 13, how many are left to read?

☐ books

SET 2 Addition and subtraction of 4- and 5-digit numbers

Populations

GLADSTONE 42 727

KALLANGUR 14 740

MAROOCHYDORE 16 598

INNISFAIL 8150

HERVEY BAY 42 392

TOOWOOMBA 90 563

Find the combined populations of these towns.

1

Maroochydore	
Hervey Bay	

2

Kallangur	
Toowoomba	

3

Gladstone	
Innisfail	

4

Toowoomba	
Maroochydore	

Calculate the difference in population between these towns:

5

Toowoomba	
Innisfail	

6

Gladstone	
Hervey Bay	

Space Constructing rectangles

Use the 5 mm dot paper to draw these rectangles.

1 2 cm × 4 cm

2 3 cm × 5 cm

3 3 cm × 3 cm

Number and Algebra

SET 3 Add and subtract fractions

1 $\frac{3}{4} - \frac{1}{4} =$

2 $\frac{6}{10} - \frac{1}{10} =$

3 $\frac{9}{10} - \frac{3}{10} =$

4 $\frac{7}{8} - \frac{5}{8} =$

5 $\frac{7}{8} - \frac{4}{8} =$

Record these answers as improper fractions and as mixed numerals.

6 $\frac{8}{10} + \frac{5}{10} =$ $\quad =$

7 $\frac{4}{5} + \frac{3}{5} =$ $\quad =$

8 $\frac{7}{10} + \frac{4}{10} =$ $\quad =$

9 $\frac{8}{10} + \frac{9}{10} =$ $\quad =$

10 $\frac{2}{4} + \frac{3}{4} =$ $\quad =$

11 $\frac{3}{4} + \frac{3}{4} + \frac{1}{4} =$ $\quad =$

12 $\frac{7}{10} + \frac{6}{10} + \frac{5}{10} =$ $\quad =$

13 $\frac{3}{5} + \frac{4}{5} + \frac{4}{5} =$ $\quad =$

14 $\frac{3}{10} + \frac{9}{10} + \frac{9}{10} =$ $\quad =$

15 $\frac{5}{8} + \frac{7}{8} + \frac{5}{8} =$ $\quad =$

SET 4 Extension

1 Average 10, 90, 170, 90

2 $\frac{3}{10} = \frac{30}{100}$ True or false?

3 4.75 m = ☐ cm

4 350 – 90 ÷ 5

5 How many sides has a decagon?

6 3.52 + 7.64

7 How many faces has a hexagonal prism?

8 $\frac{3}{4}$ of 200

9 At what temperature would water boil: 0°C, 25°C, 50°C or 100°C?

10 Write one hundred and five thousand, two hundred and twenty-six in figures.

11 37 280, 37 250, 37 220, ☐

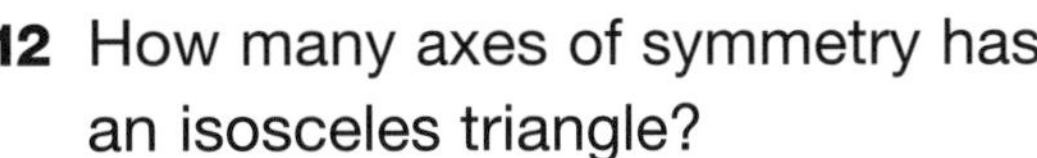

12 How many axes of symmetry has an isosceles triangle?

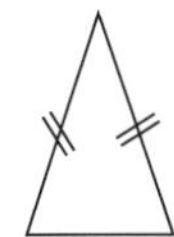

13 Which one is not equivalent: $\frac{3}{4}$, $\frac{75}{100}$, 80% or 0.75?

Mathematical Reasoning

14 Find the mean population of the following towns:

Gladstone 42 727 Kallangur 14 740
Maroochydore 16 598 Toowoomba 90 563
Hervey Bay 42 392

Measurement Kilometres

Calculate the kilometres travelled in these 4WD tours.

1 Start at A and drive to D, passing through B and C. ☐

2 Start at A and drive to D, passing through G and E. ☐

3 Start at A and drive to D, passing through B, F and E. ☐

4 Start at E and drive to D, passing through F, B and C. ☐

5 Start and end at A, passing through B, C, D, E and G. ☐

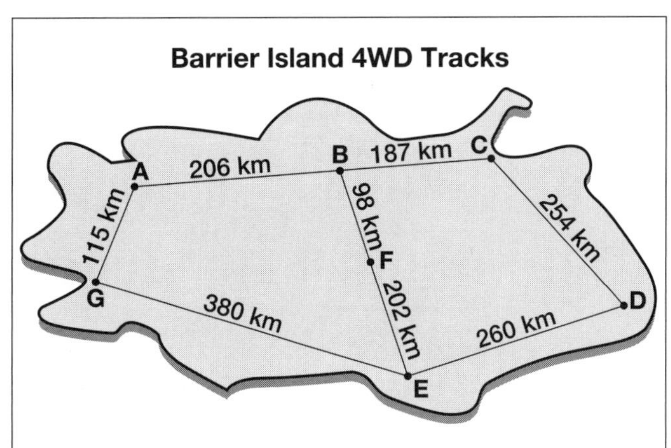

UNIT 11

Number and Algebra

SET 1 Basic

1 $25 \div 3$
2 $15 \div 3 + 10$
3 Which is the 23rd letter of the alphabet?
4 $6^2 + 3$
5 Days in 2 years
6 $42 + 19$
7 Perimeter of a square 5 cm wide
8 $9 + 500 + 40$
9 $42 - 17$
10 $21\% = 0.\square$
11 $3 \times 9 \times 0$
12 $29 \div 9$
13 Difference between 9^2 and 20
14 $8 + 600 + 70$
15 One third of 27
16 Tim has a mass of 68 kg and Jesse has a mass of 53 kg. What is the difference in their masses?
$\square$ kg

SET 2 Negative numbers

Complete these number sentences.

1 $6 + 3 - 5 - 4 =$
2 $5 + 4 - 6 - 7 =$
3 $-5 + 4 + 6 - 9 =$
4 $7 - 8 - 12 + 3 =$
5 $\$20 + \$30 - \$80 =$
6 15 kg – 20 kg + 15 kg =

Solve these problems.

7 If the temperature at 4 pm was 3°C, what was the temperature at midnight if it was 8°C lower?
8 What is the balance of Milton's bank account if he had $45 but spent $75 on bike repairs?
9 What was Navaya's final score in the board game if she was awarded 150 points but had 200 points deducted?

Statistics and Probability Line graphs

Kimberley made a graph to represent the number of customers in her shop during the day. Use the graph to answer the questions.

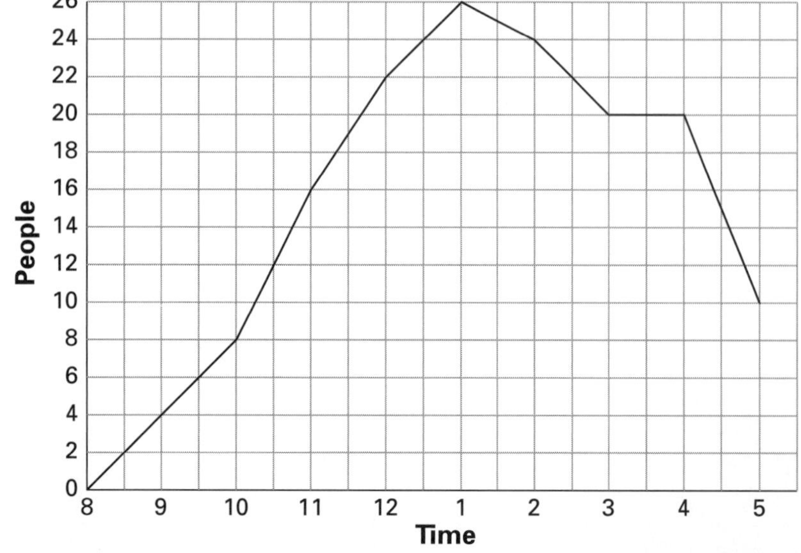

1 What was the busiest hour?
2 How many people were in the shop at 9 am?
3 Estimate the number of people in the shop at 12:30.
4 How many more people were in the shop at 4 pm compared to 5 pm?

Number and Algebra

SET 3 Unit fractions of a quantity

Find these fractions of 36 stars.

1 $\frac{1}{2}$ =

2 $\frac{1}{6}$ =

3 $\frac{1}{4}$ =

4 $\frac{1}{12}$ =

5 $\frac{1}{9}$ =

6 $\frac{1}{3}$ =

Find the fraction of each quantity.

7 $\frac{1}{3}$ of 90 =

8 $\frac{1}{5}$ of 100 =

9 $\frac{1}{4}$ of 120 =

10 $\frac{1}{5}$ of 200 =

11 $\frac{1}{5}$ of 300 =

12 $\frac{1}{4}$ of 240 =

13 $\frac{1}{8}$ of 240 =

14 $\frac{1}{5}$ of 400 =

15 $\frac{1}{8}$ of 400 =

16 $\frac{1}{5}$ of 800 =

17 $\frac{1}{8}$ of 640 =

18 $\frac{1}{10}$ of 6000 =

19 $\frac{1}{20}$ of 1000 =

20 $\frac{1}{100}$ of 2000 =

SET 4 Extension

1 Which does not fit: $\frac{1}{2}$, 0.5, $\frac{22}{40}$ or 50%?

2 Write 5:10 pm in 24-hour time.

3 What is the volume of a box with length 5 cm, width 2 cm and height 6 cm?

4 How many kilograms in 5.25 t?

5 At what temperature does ice melt?

6 Area of a rectangle with sides 40 m and 8 m

7 How much are 12 cakes at 3 for $4.75?

8 How many faces has a square pyramid?

9 x = ☐°

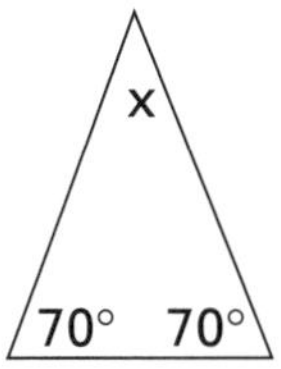

(Triangle with angles x, 70°, 70°)

10 $40 less $5.50

11 4.35 km = ☐ m

12 Round 637 428 to the nearest 10.

13 How many axes of symmetry has a regular octagon?

Mathematical Reasoning

14

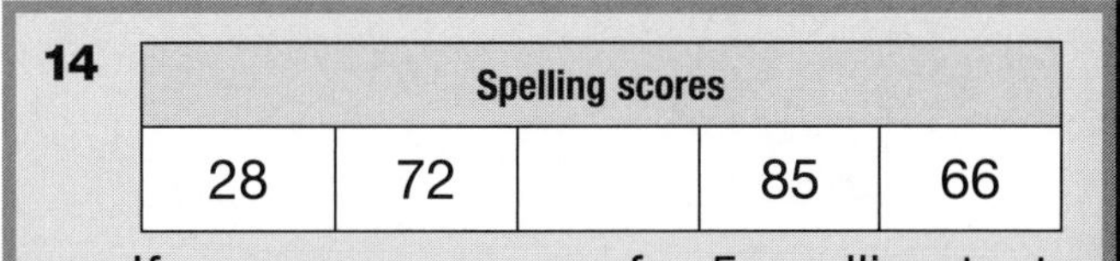

Spelling scores				
28	72		85	66

If my average score for 5 spelling tests is 65, what is the missing score?

Measurement Kilograms and tonnes

How many of each set of masses are needed to make a tonne?

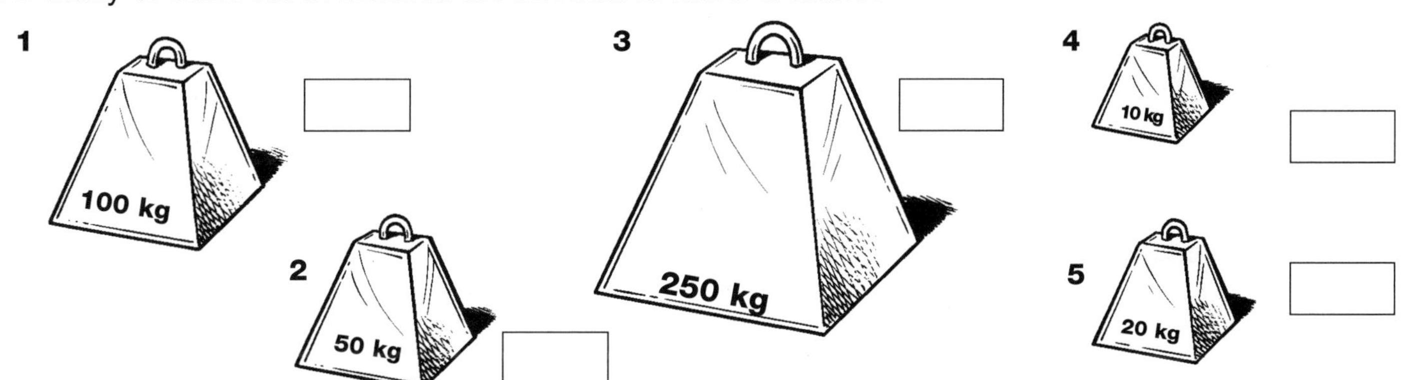

Number and Algebra

SET 1 Basic

1 150 + 30
2 20 × 6
3 600 ☐ 550 = 50
4 27 ☐ 12 = 15
5 3 ☐ 12 = 36
6 43 ÷ 6
7 107 ☐ 107 = 214
8 Product of 7 and 0
9 Sum of 127 and 20
10 0.9 as a fraction
11 Value of 6 in 364 000
12 Is 15 a prime number?
13 4 mins = 240 sec. True or false?
14 3 L and 270 mL = ☐ mL
15 $5^2 + 6$
16

SET 2 Order of operations

Use the order of operations to solve the questions.

1 $7 \times 3 + 20 =$
2 $7 \times (3 + 20) =$
3 $7 \times 5 + 9 - 6 =$
4 $35 - 3 \times 5 + 6 =$
5 $5 \times (9 + 6) + 30 =$
6 $35 + 7 - 6 \times 3 + 40 =$
7 $66 \div 6 + 7 + 40 =$
8 $200 - 7 \times 6 + 30 =$
9 $2.3 \times 2 + 5 =$
10 $(4.5 + 5.1) \div 3 =$
11 $(7.8 - 5.3) \times 4 =$
12 $\frac{1}{4} \times (18 + 6) =$
13 $0.5 \times 6 + 8 =$
14 Tom saved $497 in May, $256 in June and $387 in July towards a new TV set. How much more does he need to save if the TV costs $1497?

Space Cross sections

Next to each 3D object, draw the 2D shape that represents its cross section.

1

3
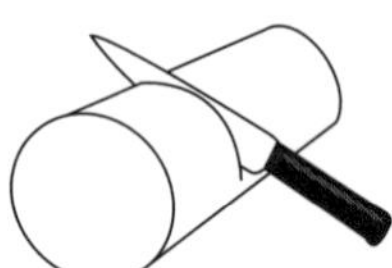

2
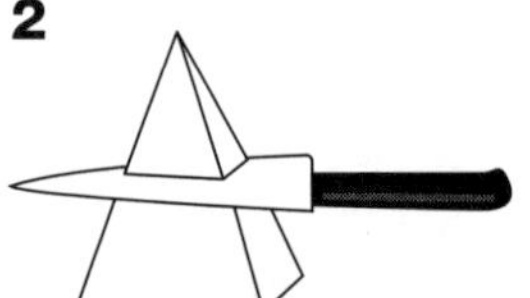

4

Number and Algebra

SET 3 Percentages, decimals and fractions

Write the missing fraction, decimal or percentage in the table.

	Fraction	Decimal	Percentage
1	$\frac{1}{10}$	0.1	%
2	$\frac{2}{10}$		20%
3	$\frac{1}{2}$		50%
4	$\frac{23}{100}$	0.23	%
5	$\frac{99}{100}$		99%
6	$\frac{1}{5}$		20%
7	$\frac{3}{4}$		75%
8	1	1.0	%

Find these simple quantities.

9 $\frac{1}{2}$ of $50

10 $\frac{1}{5}$ of $50

11 10% of $100

12 10% of $40

13 $\frac{1}{4}$ of $40

14 25% of $100

SET 4 Extension

1 37% = 0.☐

2 Order 0.31, 30%, $\frac{9}{10}$, and $\frac{1}{5}$.

3 $\frac{8}{10} + \frac{8}{10} + \frac{7}{10}$

4 How much is 3.25 m at $20 per metre?

5 How many sides has a heptagon?

6 $\frac{25}{40} = \frac{\square}{8}$

7 3.5, 4, 4.5, 5, ☐

8 How many faces has an octagonal prism?

9 What is the perimeter of a square with sides of 3.5 cm?

10 If 20 pencils cost $2.60, how much would 30 cost?

11 2 dozen lollies at 6 lollies for 75c

12 How much is 4.2 m of timber at $5 per metre?

13 How many 3 cm cubes will fit into a box measuring 6 cm in length, 3 cm in width and 6 cm in height?

Mathematical Reasoning

14 
How tall are Sally and Anna if Sally is 0.1 m taller than Anna and their combined height is 3.16 metres?

Sally: ______ Anna: ______

Statistics and Probability Likelihood

1 Four marbles, 2 red and 2 yellow, are put into a bag. Write 5 different ways that the marbles can be drawn out of the bag.

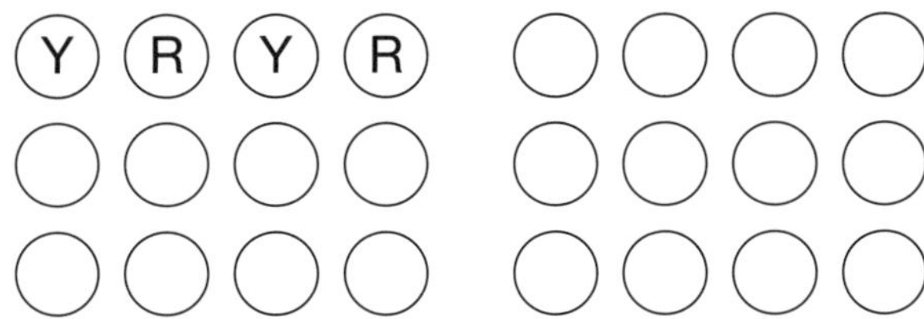

Two sets of dice have been rolled showing the total scores of 7 and 10. One example of each score is shown below.

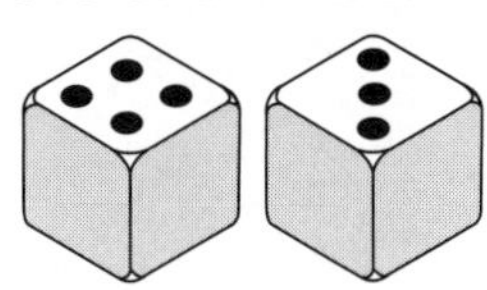
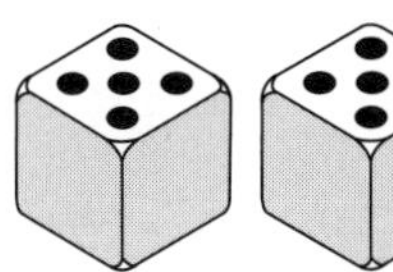

2 Which score do you think is more likely to occur? Explain why. ______________________

____

____

UNIT 13

Number and Algebra

SET 1 Basic

1 18 ☐ 12 = 30

2 100 ☐ 90 = 10

3 6^2

4 Divide 18 by 2.

5 7×7

6 24 – 8

7 18 + 21

8 26 + 17

9 Are 14 and 31 multiples of 6?

10 Is 7 a prime number?

11 Divide 13 by 6.

12 Value of 3 in 35 407

13 45c × 3

14 $3 \times 7 + 9$

15 How many 500 g bags make 2 kg?

16

Cook House scored 650 points and received 45 bonus points. How many points did Cook have?

☐ points

SET 2 Subtracting decimals

1 9.85 – 5.14

2 8.67 – 3.22

3 5.73 – 4.61

4 7.48 – 3.35

5 6.96 – 4.50

6 7.85 – 6.62

7 84.25 − 31.24

8 76.76 − 52.63

9 65.65 − 43.14

10 8.524 − 3.123

11 67.43 − 45.40

12 5.478 − 3.205

Statistics and Probability Potentially misleading data

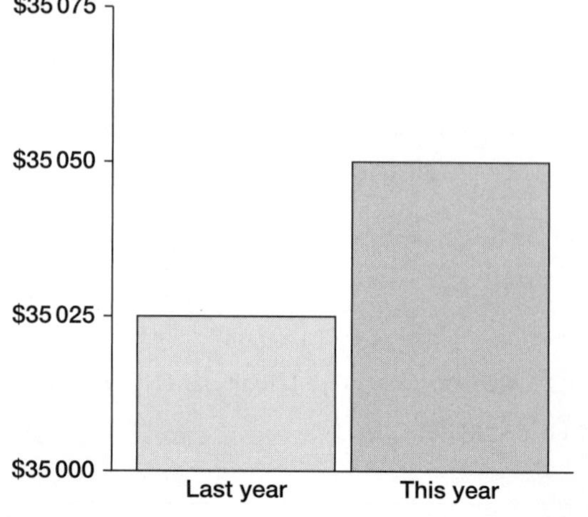

Explain how the data presented in the graph and the graph's title could be misleading.

Number and Algebra

SET 3 Expanding numbers

Expand these numbers. The first one is done for you.

1 37 423 = 30 000 + 7000 + 400 + 20 + 3

2 85 616 =

3 25 209 =

4 106 015 =

5 6 041 500 =

What number am I?

6 I am 40 000 more than 28 499

7 I am 6000 more than 148 347

8 I am 90 more than 265 268

9 I am 700 more than 70 658

10 I am 2 000 000 less than 56 812 405

Write the following numbers.

11 400 000 + 80 000 + 200 + 7 =

12 4 000 000 + 300 000 + 10 000 + 100 =

13 5 000 000 + 500 000 + 40 000 + 1000 + 40 =

SET 4 Extension

1 Perimeter of a hexagon with 3.25 m sides

2 How much are 7 pencils at 3 for $1.65?

3 How much is 7.25 kg at $20 per kilogram?

4 Order 1.53, 150%, 1.49 and $1\frac{4}{10}$.

5 Add all the odd numbers between 20 and 30.

6 Is a hectare roughly equal to 2 tennis courts or to 2 soccer fields?

7 $\frac{7}{10}$ of 2.5 m = ☐ cm

8 What is the area of a rectangle with sides 7 cm and 3.5 cm?

9 9 m – 150 cm?

10 How many cm^2 in 1 m^2?

11

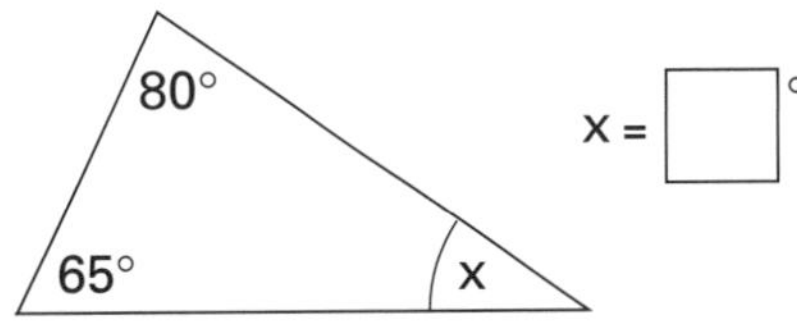

x = ☐°

12 Average $3.50, $4.50, $5.50 and $2.90.

13 How many axes of symmetry has a rhombus?

14 How many mL in 5.75 L?

15 What shape does this net make?

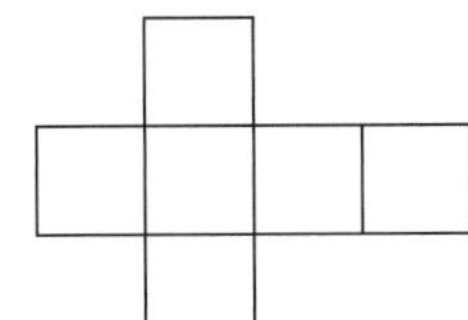

Measurement Kilometres

Each band on the radar is equal to 10 km. Estimate the distances of the submarines from the centre of the radar screen.

1 Submarine A ☐ km

2 Submarine B ☐ km

3 Submarine C ☐ km

4 Submarine D ☐ km

5 Submarine E ☐ km

6 Submarine F ☐ km

HMAS Pinafore radar

Number and Algebra

SET 1 Basic

1 75 ☐ 50 = 25

2 $4^2 + 5$

3 23 – 8

4 9 ☐ 12 = 21

5 Divide 42 by 7.

6 $1\frac{1}{4}$ hours = ☐ minutes

7 One third of 45

8 130 ÷ 10

9 Difference between $4 and $1.35

10 Value of 4 in 7043

11 Are 28 and 56 multiples of 7?

12 How many sides has an octagon?

13 How many 50c coins in $7?

14 13 km + 38 km

15

Sam saved $6 a week for 8 weeks. How much did he save?

$ ☐

SET 2 4-digit × 1-digit multiplication

1 1342 × 4

2 2245 × 3

3 4173 × 5

4 2341 × 6

5 1054 × 8

6 5183 × 7

7 3718 × 4

8 6049 × 5

9 2244 × 9

10 A company bought 8 laptops for their executive staff. If each laptop cost $5329, what was the total cost?

$ ☐

11 7950 books were sold at Christmas and 1262 at Easter. What was the total sales value if the books averaged $9 each?

$ ☐

Number and Algebra Prime or composite numbers

Write whether the numbers are prime or composite, then explain why.

	Number	Prime or composite	Explain why
1	17		
2	51		
3	39		
4	85		
5	43		

Number and Algebra

SET 3 Equivalent fractions

Write an equivalent fraction for the ones supplied.

1 $\frac{1}{2} = \frac{\square}{4}$

2 $\frac{1}{4} = \frac{\square}{8}$

3 $\frac{1}{4} = \frac{\square}{12}$

4 $\frac{1}{3} = \frac{\square}{6}$

5 $\frac{1}{2} = \frac{\square}{8}$

6 $\frac{1}{2} = \frac{\square}{10}$

7 $\frac{1}{3} = \frac{\square}{12}$

8 $\frac{1}{5} = \frac{\square}{10}$

9 $\frac{1}{5} = \frac{\square}{15}$

10 $\frac{3}{5} = \frac{\square}{10}$

11 $\frac{2}{3} = \frac{\square}{6}$

12 $\frac{3}{4} = \frac{\square}{8}$

SET 4 Extension

1 $\frac{3}{4}$ of $240

2 How many faces has a hexagonal pyramid?

3 3 dozen cakes at 45c each

4 If 16 bananas cost $2.40, how much would 20 cost?

5 Estimate 396 × 31.

6 Write 1:15 pm in 24-hour time.

7 7.4 m = $\square$ cm

8 $\frac{20}{45} = \frac{4}{\square}$

9 What fraction of 300 is 25?

10 How many 1.4 m lengths can be cut from a 7 m piece?

11 How many 300 g bags in $4\frac{1}{2}$ kg?

12 How many 6 cm squares are needed to cover a rectangle with sides 12 cm and 24 cm?

13 How much is 2.25 kg at $4.40 per kilogram?

Mathematical Reasoning

14 The park and the preschool each occupy 10 000 square metres of land but are different shapes. Which one has the larger perimeter if the preschool is a square with 100 metre sides, and the park is a rectangle 200 metres long and 50 metres wide?

Space Reflections

Use the dot paper to reflect each shape.

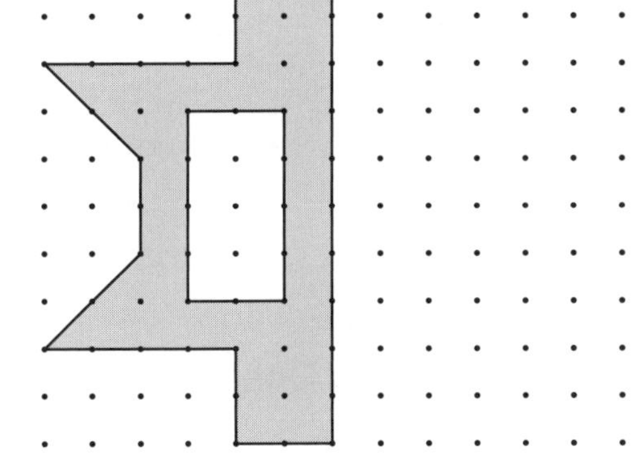

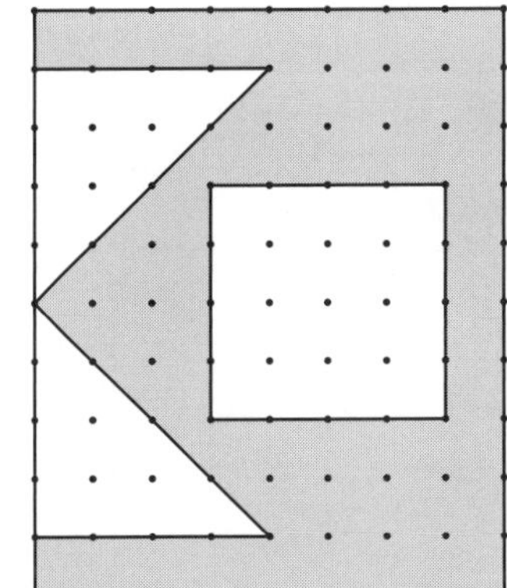

Number and Algebra

SET 1 Basic

1 One position after 21st

2 $9 + 2\frac{1}{2}$

3 $16 \div 5$

4 18 ☐ 3 = 6

5 40 mm + 5 cm = ☐ mm

6 Next ordinal number after 3rd

7 What is the date one week after Anzac Day?

8 Difference between 16 and 3^2

9 $\frac{1}{2}$ of $0.70

10 $\$0.24 \times 2$

11 $200 \div 10$

12 Square 5.

13 Cost of ten 65c stamps

14 8^2

15

SET 2 Comparing and ordering fractions

1 Draw a line to your estimate of the place of each fraction on the number line.

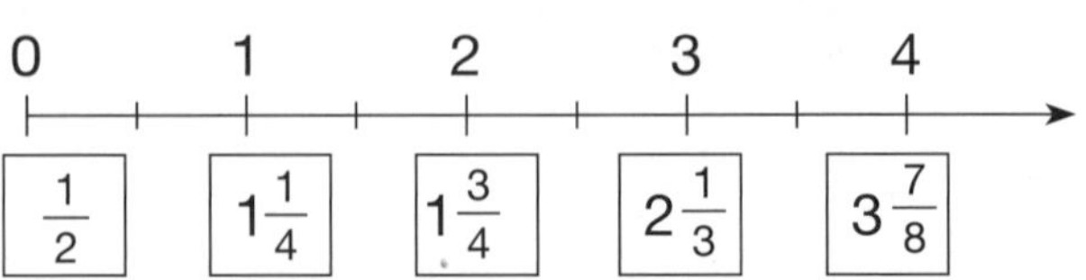

2 Label each place on the number line.

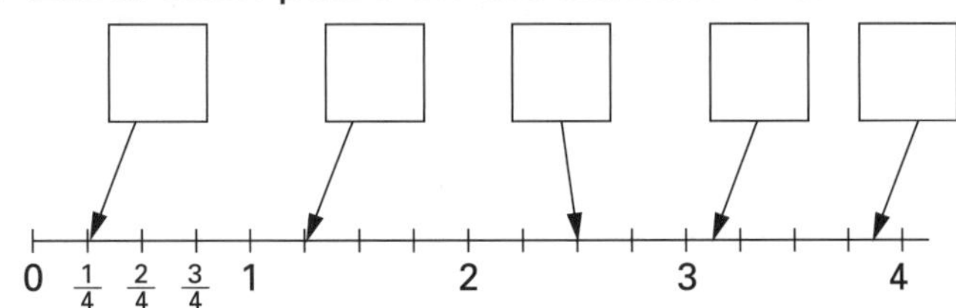

Continue the sequences.

3	$3\frac{1}{2}$	$4\frac{1}{2}$	$5\frac{1}{2}$			
4	$7\frac{1}{4}$	$7\frac{1}{2}$	$7\frac{3}{4}$			
5	$8\frac{1}{3}$	8	$7\frac{2}{3}$			
6	$6\frac{1}{4}$	$6\frac{3}{4}$	$7\frac{1}{4}$			
7	7	$7\frac{3}{4}$	$8\frac{1}{2}$			
8	$5\frac{3}{5}$	$6\frac{1}{5}$	$6\frac{4}{5}$			
9	$7\frac{7}{10}$	$8\frac{3}{10}$	$8\frac{9}{10}$			

Space Problem solving using scale

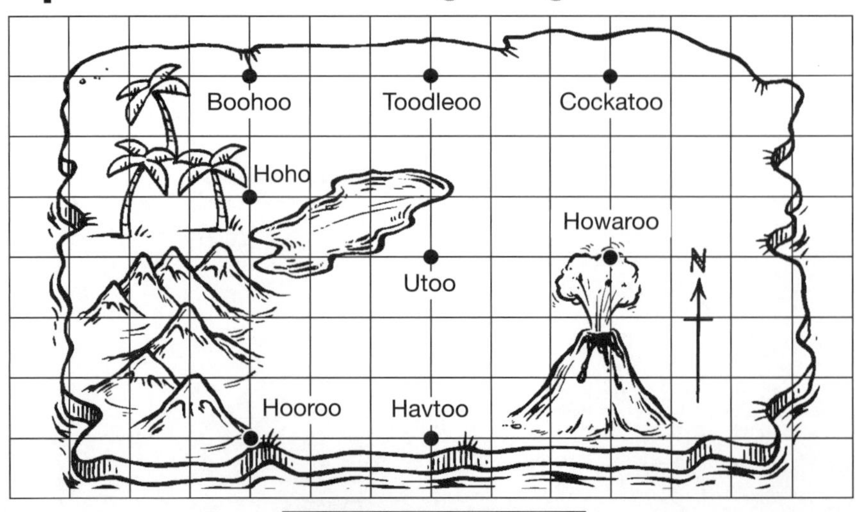

Scale: 8 mm = 3 km

Use the scale to calculate the distance between:

1 Boohoo and Cockatoo

2 Toodleoo and Utoo

3 Hooroo and Hoho

If a taxi averages $3.50 per kilometre, how much would each taxi trip cost?

4 Boohoo to Cockatoo $_____

5 Toodleoo to Utoo $_____

6 Hooroo to Hoho $_____

Number and Algebra

SET 3 Rounding

Round each number to the nearest 1000 to estimate an answer, then calculate the exact answer.

	Question	Estimate	Exact
1	1891 + 4163	2000 + 4000 = 6000	6054
2	8761 – 2845		
3	6117 + 3922		
4	3618 + 2386		
5	7762 + 1224		
6	5681 – 3749		

Round to the nearest 1000 to estimate the product.

7 $4956 \times 4 \approx$

8 $5708 \times 7 \approx$

9 $9119 \times 9 \approx$

10 $6040 \times 5 \approx$

11 $7697 \times 2 \approx$

12 $3852 \times 8 \approx$

13 $8127 \times 3 \approx$

SET 4 Extension

1 $19.15 = [ ] c

2 Share $16.50 among 3.

3 If tomatoes are $1.80 a kg, how much would 4500 g cost?

4 $36 + 17 + 9 \times 7$

5 $46 + 7 \times 8 + 3$

6 If Mum drove 260 km in three hours, was her average speed about 90 km/h or 26 km/h?

7 How many metres in 5.25 km?

8 Change from $10 if I bought three bracelets at a cost of $2.70 each

9 Arrange in ascending order: 5.0, 0.5, 5^2, 15.0

10 7.3 kg at $20 per kilogram

11 How many degrees between south and north-west?

Mathematical Reasoning

12 Yolande said, 'If you double the sides of a cube you double its volume.'

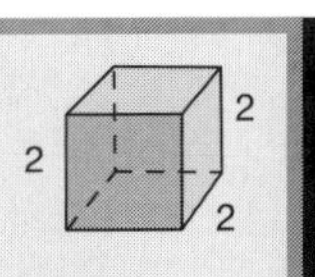

Test Yolande's theory and explain your results.

____

____

Statistics and Probability Chance from 0 to 1

Rate the likelihood of spinning each colour on the spinner using the range 0–1.

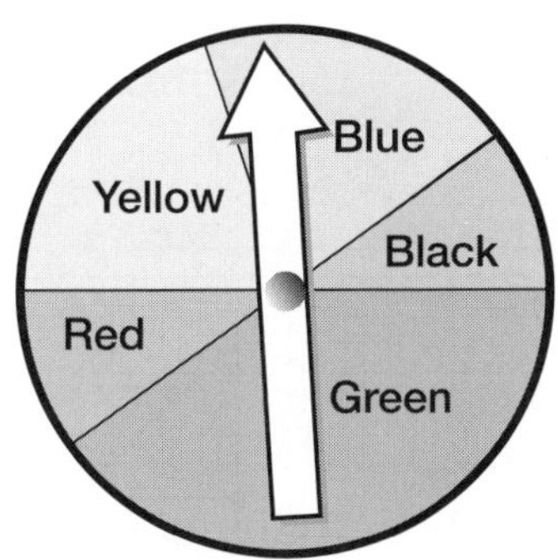

	Colour	Likelihood
1	Yellow	
2	Blue	
3	Black	
4	Green	
5	Red	

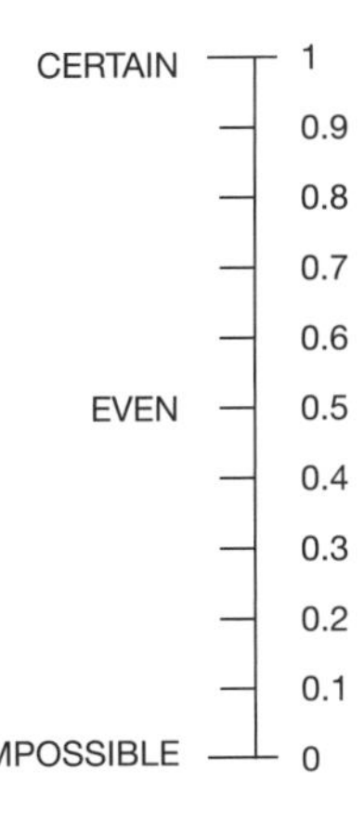

Number and Algebra

SET 1 Basic

1 9^2

2 42 + 19

3 76 – 38

4 62c × 4

5 $2\frac{3}{4}$ hours = ☐ minutes

6 2 L and 25 mL = ☐ mL

7 Which is larger, \$2.50 × 10 or \$250?

8 Add $\frac{6}{10}$ to 3.37.

9 Cost of nine 60c stamps

10 $\frac{19}{100}$ = 0.☐

11 Difference between 7^2 and 19

12 80 + 1000 + 400 + 3

13 Seconds in 10 minutes

14 (18 ÷ 9) + 10

15 One fifth of 45

16 A bricklayer lays 2 bricks a minute. How many will he lay in 45 minutes? ☐ bricks

SET 2 Division with fractional remainders

1 $3\overline{)6129}$

2 $5\overline{)1845}$

3 $7\overline{)7861}$

4 $4\overline{)2632}$

5 $6\overline{)2322}$

6 $10\overline{)6510}$

Record the remainders as fractions.

7 $5\overline{)6883}$

8 $4\overline{)7302}$

9 $3\overline{)5857}$

10 $6\overline{)7394}$

11 $8\overline{)6603}$

12 $5\overline{)7264}$

13 A box of 1456 lollies was sorted into 4 piles. How many in each pile?

14 725 Christmas lights were put into 5 packets. How many lights were in each packet?

Number and Algebra Geometric patterns

Draw the next set of hexagons in the box.

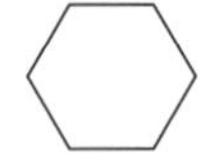

1 Complete and extend the table to record the number of sides needed to make the pattern of hexagons.

Hexagons	1	2	3	4	5	6	7
Sides							

2 Use the table to state how many sides there would be for 10 hexagons ______________

Number and Algebra

SET 3 Adding fractions—related denominators

1 whole											
$\frac{1}{2}$						$\frac{2}{2}$					
$\frac{1}{3}$				$\frac{2}{3}$				$\frac{3}{3}$			
$\frac{1}{6}$		$\frac{2}{6}$		$\frac{3}{6}$		$\frac{4}{6}$		$\frac{5}{6}$		$\frac{6}{6}$	
$\frac{1}{12}$	$\frac{2}{12}$	$\frac{3}{12}$	$\frac{4}{12}$	$\frac{5}{12}$	$\frac{6}{12}$	$\frac{7}{12}$	$\frac{8}{12}$	$\frac{9}{12}$	$\frac{10}{12}$	$\frac{11}{12}$	$\frac{12}{12}$

1. $\frac{1}{3} + \frac{1}{6} =$
2. $\frac{1}{6} + \frac{1}{12} =$
3. $\frac{1}{3} + \frac{1}{12} =$
4. $\frac{1}{2} + \frac{1}{6} =$
5. $\frac{1}{2} + \frac{1}{12} =$

Record these answers as improper fractions and as mixed numerals.

6. $\frac{5}{6} + \frac{4}{6} = \quad =$
7. $\frac{4}{6} + \frac{3}{6} = \quad =$
8. $\frac{7}{12} + \frac{4}{6} = \quad =$
9. $\frac{9}{12} + \frac{5}{6} = \quad =$
10. $\frac{11}{12} + \frac{1}{6} = \quad =$

SET 4 Extension

1. Round and estimate: \$2.95 × 18.
2. $\frac{4}{5}$ of \$300
3. $\frac{4}{8} + \frac{1}{8} + \frac{7}{8}$
4. How much are 6 books at 5 for \$7.50?
5. 9.49 – 4.61
6. 20% of \$150
7. Perimeter of a decagon with 7.5 cm sides
8. Half of $5\frac{1}{2}$ dozen
9. 60% of 90
10. How many fifths in 7.4?
11. Order $1\frac{3}{5}$, 1.61, 1.5 and 1.
12. \$50 – \$39.90
13. $(8^2 - 7^2) \times 2$
14. 70% of 1 tonne
15. What fraction of 3 km is 200 m?
16. Volume of a box with length 7 cm, width 4 cm and height 8 cm?
17. Complete the grid to convert these improper fractions to mixed numbers.

$\frac{3}{2}$	$\frac{9}{8}$	$\frac{7}{5}$	$\frac{6}{4}$	$\frac{5}{3}$	$\frac{15}{10}$	$\frac{5}{4}$	$\frac{7}{3}$
$1\frac{1}{2}$	$1\frac{1}{8}$						

Measurement Calculating volume

What is the volume of these boxes?

Volume = Length × Width × Height

1

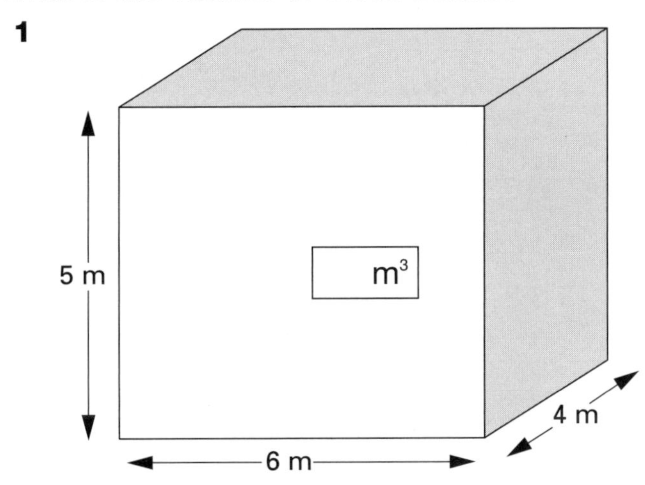

2

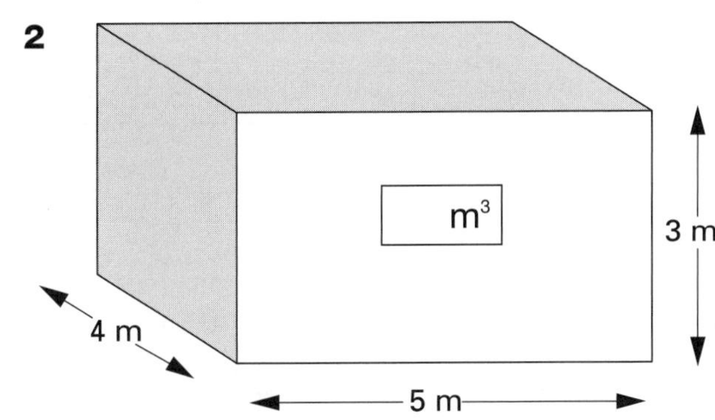

UNIT 17

Number and Algebra

SET 1 Basic

1 11 + 31

2 7 × 9

3 35 – 6

4 28 ÷ 7

5 250 ☐ 2 = 500

6 500 ☐ 10 = 50

7 Product of 8 and 10

8 Divide 88 by 11.

9 Sum of 88 and 11

10 $5^2 + 7$

11 13, 26, 39, ☐

12 Is 43 a prime number?

13 Are 200 and 16 multiples of 6?

14 5 hundreds + 1387

15 How many 500 g packs are needed to fill 2 kg?

16

How many 200 g bags can be filled from a 2 kg bucket?
☐ bags

SET 2 Extended multiplication

1 25 × 16

2 31 × 15

3 28 × 45

4 76 × 48

5 80 × 26

6 75 × 33

7 382 × 27

8 251 × 72

9 462 × 54

10 474 × 43

11 586 × 29

12 648 × 62

Number and Algebra Adding and subtracting decimals

1 $546.23 + 49.05 + 372.61

2 $785.95 + 231.29 + 502.48

3 $957.85 – 378.90

4 $429.66 – 204.80

5 6599.25 kg + 2141.25 + 4040.40

6 25.954 km + 36.612 + 52.156

7 3621.49 m – 2050.66

8 95.58 kg – 27.345

Number and Algebra

SET 3 Number patterns

Complete the tables to show how much money each person earned during the day.

1 Melissa: $8 an hour.

Hours	1	2	3	4	5	6	7
$	8	16	24				

2 Caleb: $12 an hour.

Hours	1	2	3	4	5	6	7
$	12						

3 Joseph: $20 an hour.

Hours	1	2	3	4	5	6	7
$							

4 Kate: $4.50 an hour.

Hours	1	2	3	4	5	6	7
$							

Mathematical Reasoning

5 Create your own table to show how much the manager will receive for 7 hours work.

Hours	1	2	3	4	5	6	7
$							

SET 4 Extension

1 15.25 m = ☐ cm

2 How many months in 3.75 years?

3 What fraction of 2 m is 25 cm?

4 Add all even numbers between 31 and 37.

5 How many faces has a pentagonal pyramid?

6 Is 57 a prime number?

7 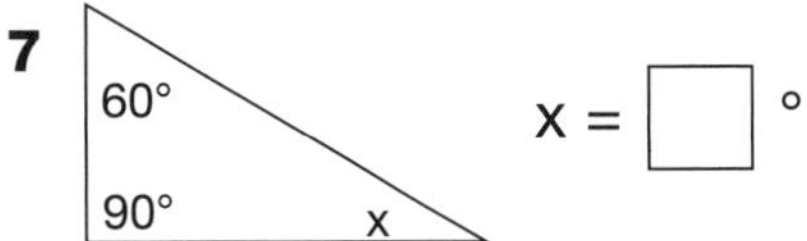

8 A hectare could be viewed as a square with 100 m sides. True or false?

9 If a 250 L tank is $\frac{7}{10}$ full, how much more could it hold?

10 Write 11:50 pm in 24-hour time.

11 Which is larger, 77.7 or $77\frac{7}{100}$?

12 Decrease 1 426 231 by 400 000.

13 7.1 L = ☐ mL

14 How many axes of symmetry has a trapezium?

15 What object does this net make?

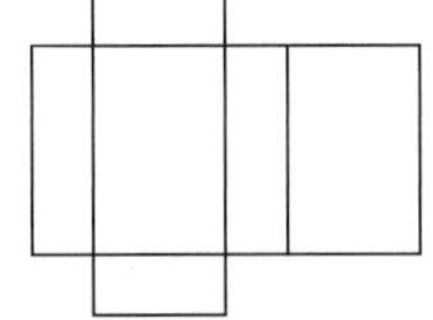

16 How many 2 cm cubes could fit into a box with length 8 cm, width 4 cm and height 10 cm?

Measurement Tonnes and kilograms

How many of each item can be placed in the container?

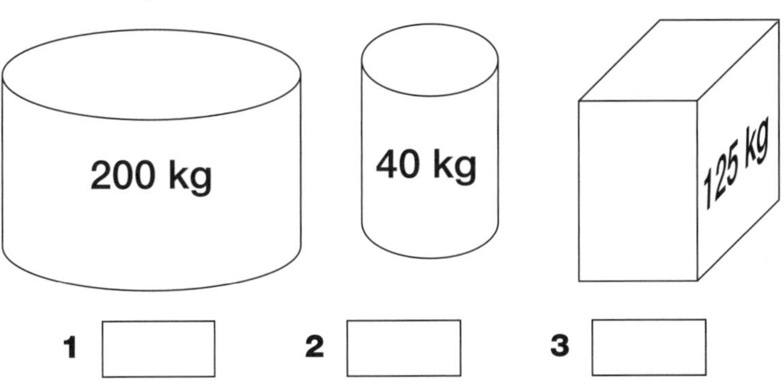

Number and Algebra

SET 1 Basic

1 6×8
2 7×9
3 $60 - 15$
4 16 ☐ 5 = 3 r 1
5 $48 \div 8$
6 22 ☐ 100 = 122
7 Is 17 a prime number?
8 Difference between 46 and 24
9 $(3 \times 7) + (3 \times 3)$
10 Factors of 18
11 $37.45 = ☐ c
12 How many sides has a quadrilateral?
13 Minutes in $3\frac{1}{2}$ hours
14 7 m and 1 cm = ☐ cm
15 Value of 8 in 13 784
16 If the tickets cost $28 each, how much will three tickets cost?
$ ☐

SET 2 Fractions of a quantity

Find the fraction of these numbers.

1 $\frac{1}{4}$ of 20
2 $\frac{1}{5}$ of 30
3 $\frac{1}{2}$ of 12
4 $\frac{1}{3}$ of 9
5 $\frac{2}{5}$ of 15
6 $\frac{1}{3}$ of 42
7 $\frac{1}{4}$ of 64
8 $\frac{1}{6}$ of 72
9 $\frac{1}{3}$ of 75
10 $\frac{1}{5}$ of 150

True or false?

11 $\frac{1}{2}$ of 30 < $\frac{1}{4}$ of 20
12 $\frac{1}{5}$ of 30 > $\frac{1}{3}$ of 24
13 $\frac{1}{3}$ of 18 < $\frac{1}{4}$ of 32
14 $\frac{1}{3}$ of 24 < $\frac{1}{5}$ of 60
15 $\frac{1}{4}$ of 60 = $\frac{1}{3}$ of 36

Mathematical Reasoning

16 A $8  B $6  C $12 D $4 

Which book was the cheapest to buy if I paid:
$\frac{1}{2}$ price for Book A
$\frac{1}{3}$ price for Book B
$\frac{1}{4}$ price for Book C
Full price for Book D

Space Translating shapes

Translate the shapes.

1 To the right

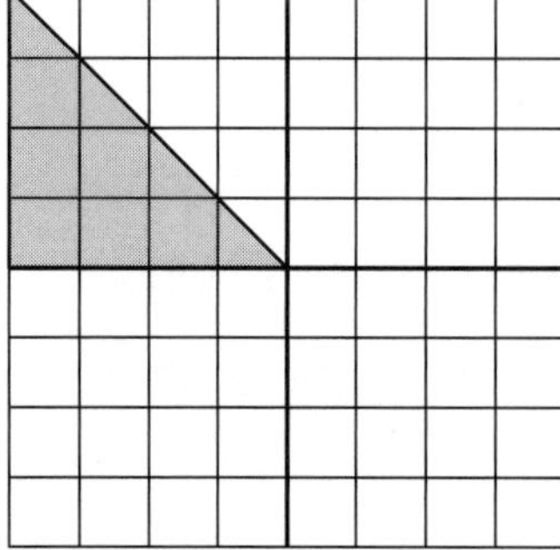

2 Directly below

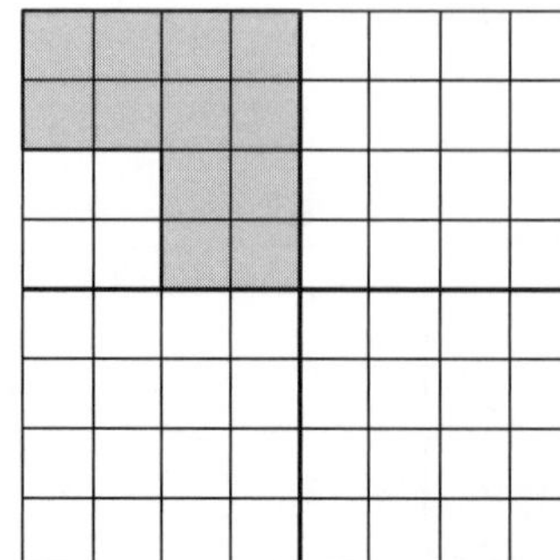

3 Directly above

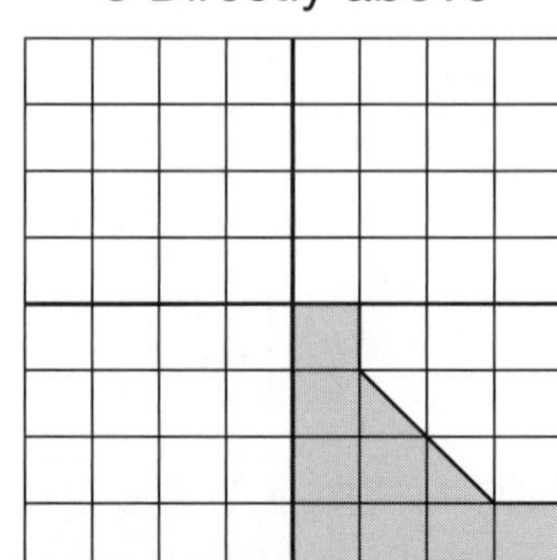

Number and Algebra

SET 3 Applying addition to solve problems

1. 3704 + 2204 = ______

2. 3628 + 4719 = ______

3. 3876 + 666 = ______

4. 4720 + 7089 = ______

5. 3625 + 703 + 80 + 924 = ______

6. 3078 + 609 + 860 + 7436 = ______

7. Michael saved $1206 in February, $986 in March and $1807 in April. What was his total for the 3 months?

8. How much is Stella's set of 4 paintings worth if they were valued at $4125, $3250, $1899 and $5455?

SET 4 Extension

1. $\frac{1}{4}$ of a day = ☐ minutes
2. 0.25 of $200
3. 25% of $36
4. 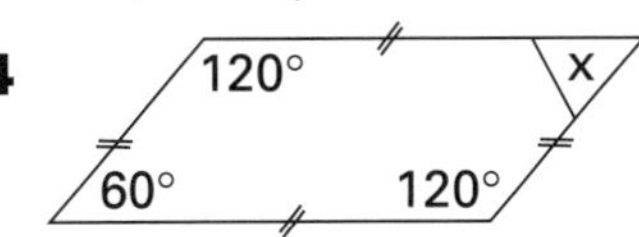

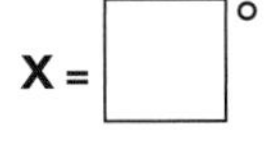

5. $(7^2 - 4^2) + 37 - 16$
6. Value of 9 in 3 936 426
7. How many 750 mL buckets are needed to fill a 6 L container?
8. John travelled 630 km in 7 hours. What was his average speed?
9. Days from 16 August to 17 September
10. Average 4.5, 6.3, 7.2, 6
11. Value of 6 in 6 342 937
12. Write 10:57 pm in 24-hour time.
13. How much is 8.75 kg at $16 per kilogram?
14. How many axes of symmetry has a semicircle?
15. Tom is 1.75 m tall. How tall is Joe if he is $\frac{4}{5}$ of Tom's height?

Measurement Cubic centimetres and millilitres

Complete the comparison chart to show equivalence.

Cubic centimetres	Millilitres
3 cm^3	mL
6 cm^3	mL
cm^3	12 mL
cm^3	24 mL
36 cm^3	mL
24 cm^3	mL

Cubic centimetres	Millilitres	Litres
cm^3	200 mL	
250 cm^3	mL	
cm^3	400 mL	
1000 cm^3	mL	1 L
cm^3	2000 mL	2 L
3000 cm^3	mL	3 L

Number and Algebra

SET 1 Basic

1 7 × 8

2 95 – 36

3 Grams in 4 kg

4 33 + 22 + 11

5 63 ☐ 9 = 7

6 96 ÷ 8

7 120 ÷ 2

8 Divide 63 by 7

9 $576.31 = ☐ c

10 $1.20 × 4

11 Value of 7 in 33 706

12 Are 18 and 42 multiples of 6?

13 Does a pentagon have 6 sides?

14 How many hours from 9 am to 2 pm?

15 How much change would I get from $2 if I bought nine 15c lollies?

$ 

SET 2 Operations with decimals

1 3.27 + 2.65

2 8.76 – 3.54

3 13.45 + 4.54 + 6.55

4 $14.45 × 2

5 $15 – $12.45

6 ($5.65 + $3.35) ÷ 3

7 5.787 km + 3.202 km

8 $100 – ($25.50 + $20.10)

9 Share $48.56 among 4

10 Cost of 6 tickets at $3.20 each

11 $15.65 + ______ = $20.20

12 7.536 metres ÷ 6

Mathematical Reasoning

Colonel Saunders recorded the price of hamburgers in 5 countries.

USA	England	Taiwan	China	Russia
$3.00	$3.40	$3.20	$1.45	$2.45

Work out the difference in price if you were buying hamburgers in these countries.

13	USA and China	
14	England and Russia	
15	Taiwan and China	
16	England and China	

Measurement 24-hour time

Display each 24-hour time on the clock faces and express each time in digital form underneath.

1 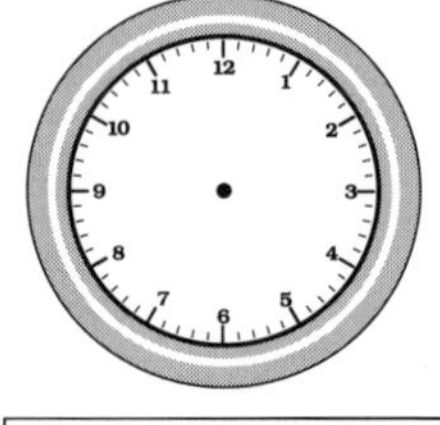

2118

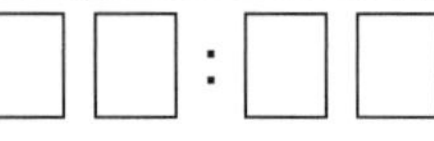

2 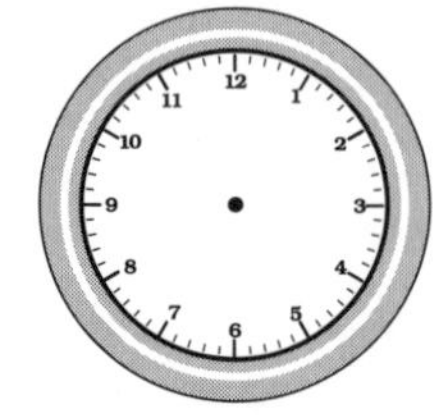

0405

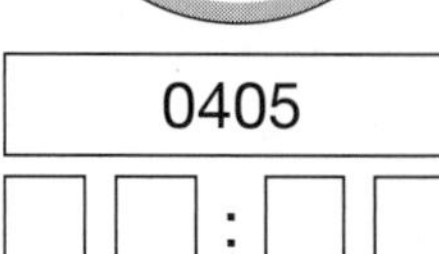

3 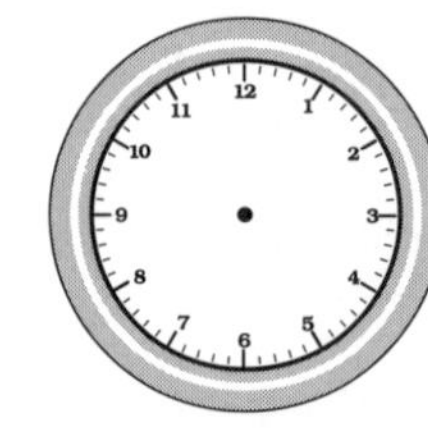

1713

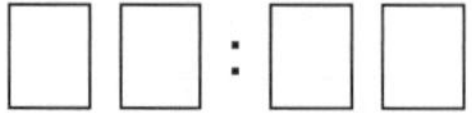

4 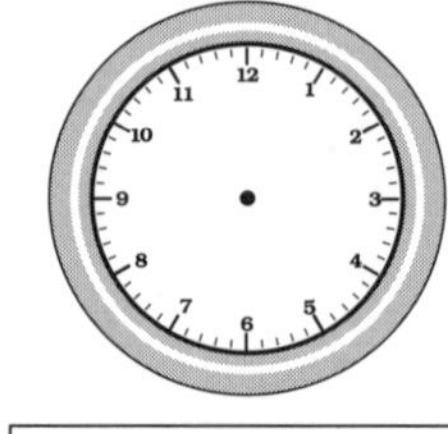

1344

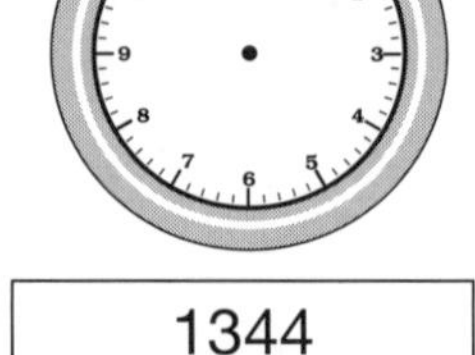

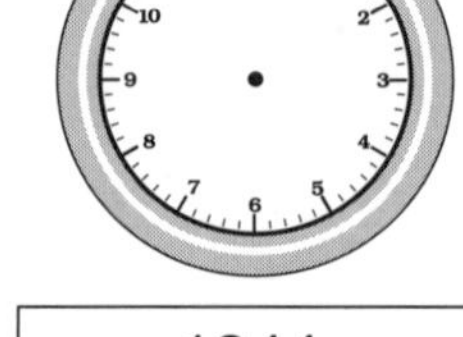

5 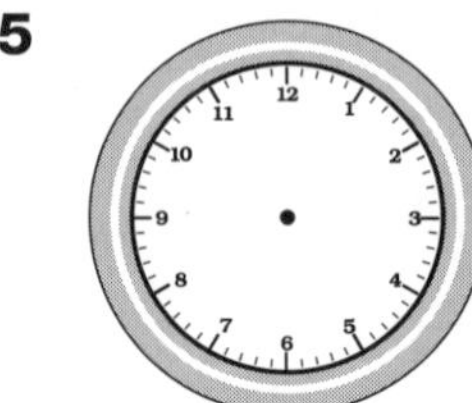

1142

Number and Algebra

SET 3 Decimal multiplication/place value

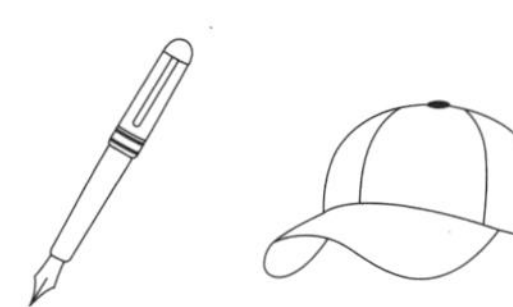

Pen	Cap	Shoes	Game
$1.24	$15.23	$34.55	$35.50

Calculate the costs. You may need to do the working on scrap paper.

1 6 pens ______________________

2 3 caps ______________________

3 5 caps ______________________

4 5 shoes ______________________

5 7 shoes ______________________

6 4 games ______________________

SET 4 Extension

1 2300 hrs = ☐ pm

2 What is the third angle of a triangle if the other angles are 48° and 29°?

3 How many hours in 3.75 days?

4 Which is greater, 25% of 1000 or 40% of 600?

5 What is the value of 9 in 28.194?

6 Round 158 960 to the nearest hundred.

7 A cook dropped $\frac{3}{8}$ of 4 dozen eggs. How many eggs were left?

8 Complete this sequence: 1, 4, 9, ☐, 25, ☐, 49

9 Estimate 298 × 21

10 How much is 5.3 m at $7 per metre?

11 Reduce 258 008 by 10 000

12 20% of $750

13 $\frac{24}{100} = \frac{\square}{25} =$

14 Which is larger, 510 mm or 0.5 metres?

15 $\frac{1}{4} + \frac{1}{2} + \frac{7}{8} =$

Mathematical Reasoning

16 Mr West had 400 sheep. If 20% of them died during a drought and 10% of the rest were sold, how many were left?

Statistics and Probability Graphing ordered pairs

Find the rule to complete the sequence, then plot each ordered pair on the graph. It has been started for you.

1

0	1	2	3	4	5
2	3				

+2

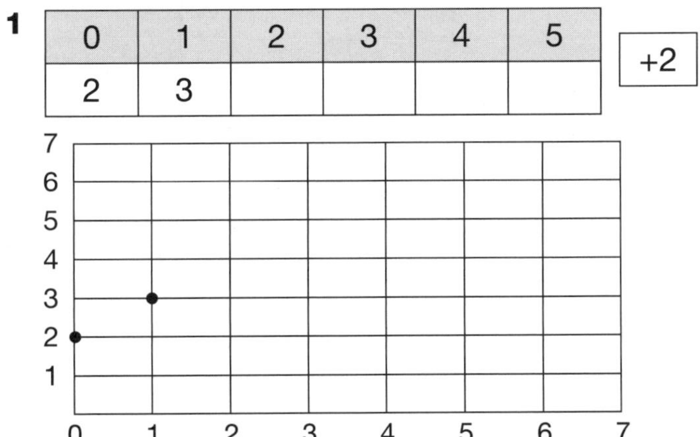

2

6	5	4	3	2	1
5					

−1

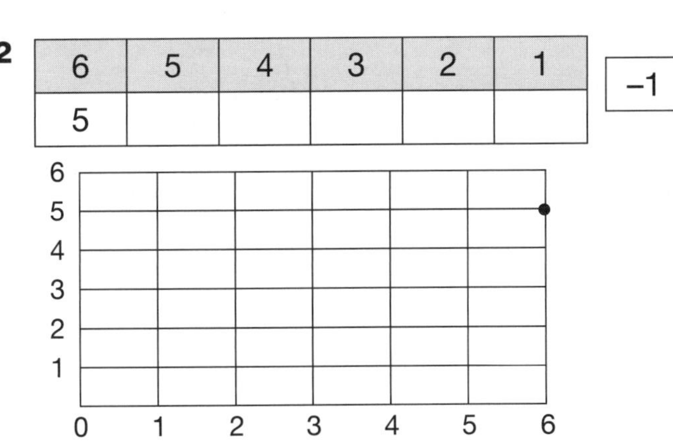

Number and Algebra

SET 1 Basic

1 85 – 29

2 Degrees in half a circle

3 30 ÷ 6

4 72 ☐ 28 = 100

5 Subtract 38 from 49.

6 2 + 40 + 1000 + 600

7 How many 20c coins in $10?

8 Factors of 21

9 $10 – $3.75

10 Add 6 to the product of 3 and 2.

11 1997 + 35

12 Value of 6 in 9.06

13 121 ☐ 11 = 11

14 16 kg and 25 g = ☐ g

15

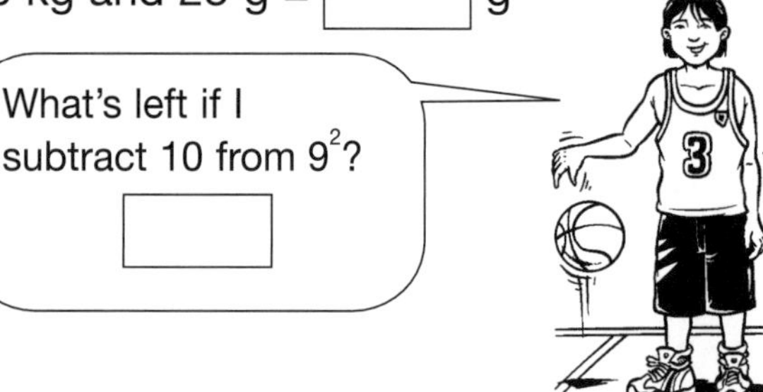

SET 2 Multiplication is commutative

Multiply these numbers in any order so that it is easier for you.

1 25 × 31 × 4

2 50 × 18 × 2

3 5 × 86 × 2

4 20 × 45 × 5

5 50 × 15 × 2

6 10 × 3 × 7

7 7 × 100 × 2

8 $4.50 × 11 × 2

9 $2.50 × 40 × 2

10 $1.25 × 12 × 4

11 How many grams of coffee are there on the shelves if each of the 3 shelves have 4 250 g jars of coffee on them?

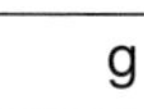

☐ g

Space Cross sections

Draw lines to connect each 3D object with its cross section.

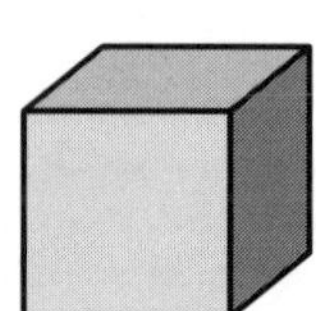
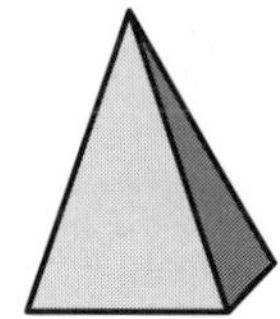
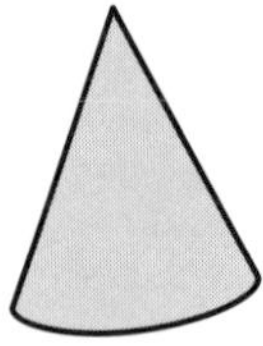
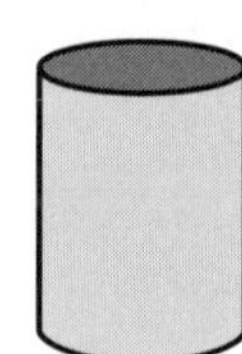
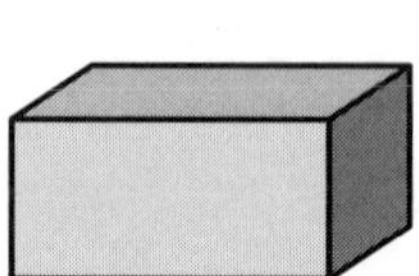

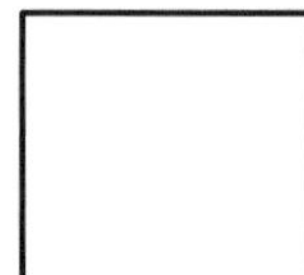
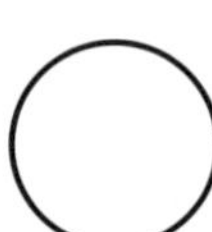
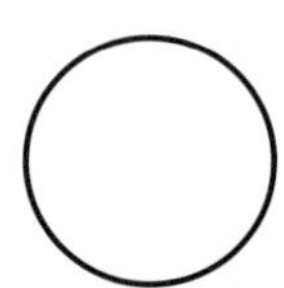

Number and Algebra

SET 3 Money problems

Write a number sentence to describe each problem then solve it.

1 Peter saved $150 per week for 5 weeks and $175 per week for 6 weeks. How much did he save?

2 Heather bought 5 cans of soup for $3.45, 7 packets of biscuits for $2.75 and 3 L of cordial at $4.50 per litre. How much did she spend?

3 John saves $\frac{1}{4}$ of his $640 weekly wage. How much will he have saved in 13 weeks?

4 Tom's medicine costs $500 for 100 mL. How much does it cost per day if Tom has to take 25 mL daily?

5

SET 4 Extension

1 0.85 = ☐ %

2 $\frac{4}{10} + \frac{3}{10} + \frac{57}{100}$

3 How much are 24 stickers at 6 for 55c?

4 How much is 5.2 m of timber at $6 per metre?

5 How many minutes from 6:45 pm to 10:07 pm?

6 If a tap drips 2 L every 10 minutes, how many litres would it drip in a day?

7 Round 410 601 to the nearest thousand.

8 Add the prime numbers between 4 and 15.

9 How many litres of oil could I buy with $18 if oil costs $2.25 per litre?

10 20% of 90

11 If Mum bought 40 L of petrol and paid 85c per litre, how much did it cost her?

Mathematical Reasoning

12 Sally's average batting score over 5 games was 16. Give a set of 5 scores that have a mean score of 16.

☐ ☐ ☐ ☐ ☐

Statistics and Probability Side-by-side column graphs

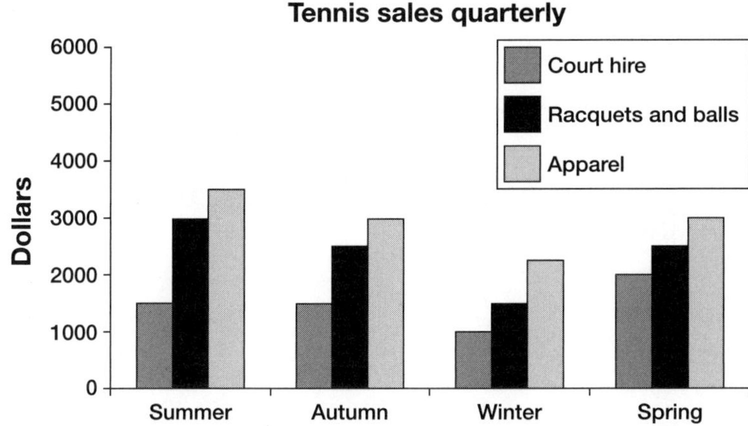

Which season was best for:

1 Apparel sales? ______

2 Racquet and ball sales? ______

3 Court hire? ______

4 How much was the total sales for apparel during winter? ______

Number and Algebra

SET 1 Basic

1 Half of 8^2

2 10:15 + 30 mins

3 420 – 170

4 How many quarters in $2\frac{1}{4}$?

5 (6 + 3) × 7

6 \$9.51 = ☐ c

7 10°C less than boiling point

8 2.1 × 7

9 2.4 + 3.5

10 94 + 72 + 6

11 Which is larger, 5^2 or 30?

12 Product of 9 and 9

13 29 km + 38 km

14 Difference between 3 and 40

15

SET 2 Spreadsheets

Year 6 are planning a farewell party. They have raised \$1998 but have some expenses to deduct.

1 Complete the spreadsheet to show the balance in their account after each expense.

	Year 6 farewell party			
	A	B	C	
1	**Category**	**Expenditure**	**Balance**	
2	Opening			\$1998
3	Prizes	\$250	= C2 – B3	\$
4	Drinks	\$265	= C3 – B4	\$
5	Decorations	\$240	= C4 – B5	\$
6	Food	\$380	= C5 – B6	\$
7	Costumes	\$235	= C6 – B7	\$
8	Printing	\$130	= C7 – B8	\$
9	Gifts	\$230	= C8 – B9	\$
10	Music	\$260	= C9 – B10	\$

2 What was the total amount spent on food and drink?

\$ 

3 What was the total amount spent on gifts, prizes and costumes?

\$ ☐

Measurement Area

Use the scale to calculate the cost of carpeting the rooms.

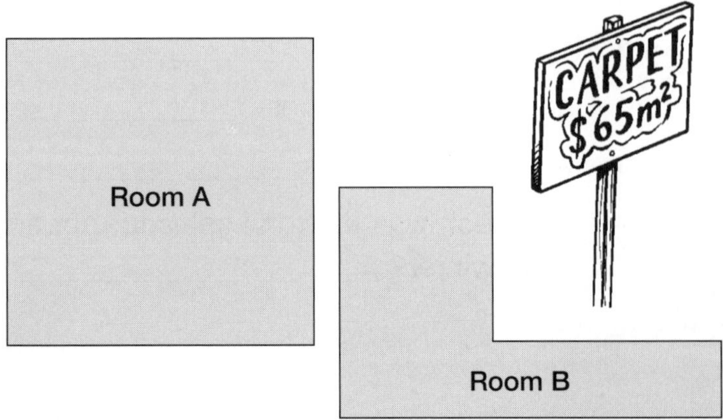

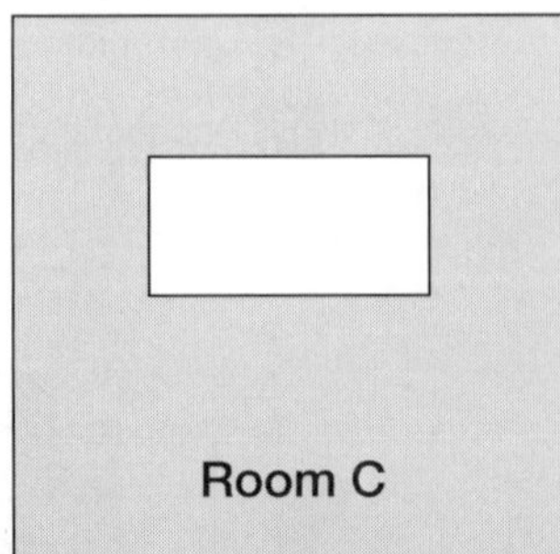

Room	Area	Cost
A		
B		
C		

Scale 1 cm = 1 m

Number and Algebra

SET 3 Equivalent fractions

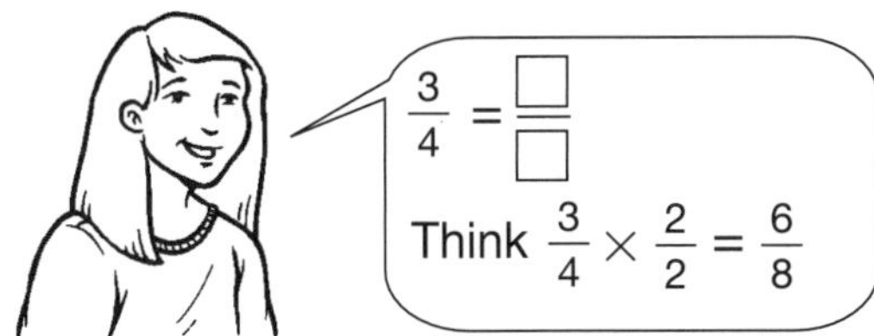

Make an equivalent fraction by multiplying the numerator and denominator by the same number.

1 $\frac{1}{5} \times \frac{2}{2} =$

2 $\frac{1}{10} \times \frac{3}{3} =$

3 $\frac{1}{6} \times \frac{4}{4} =$

4 $\frac{2}{5} \times \frac{3}{3} =$

5 $\frac{3}{4} \times \frac{5}{5} =$

6 $\frac{2}{3} \times \frac{4}{4} =$

Divide the numerator and denominator by the same number to make equivalent fractions.

7 $\frac{4}{10} \div \frac{2}{2} =$

8 $\frac{9}{15} \div \frac{3}{3} =$

9 $\frac{8}{12} \div \frac{4}{4} =$

10 $\frac{10}{15} \div \frac{5}{5} =$

11 $\frac{6}{10} \div \frac{2}{2} =$

SET 4 Extension

1 Round and estimate: 18.9 × 4.3.

2 \$8.42 × 10

3 Average of 270, 360 and 540

4 15% of \$90

5 Estimate an answer to 39 × 306.

6 72.4 + 49.27

7 $\frac{1}{4}$ kg at \$5.20 a kg

8 Order 137%, 1.4, $1\frac{3}{10}$ and 1.45.

9 (4.5 + 4.5) × 6

10 Round 826 900 to the nearest 1000.

11 Write 127% as a decimal.

12 How many seconds in 3 hours?

13 Sum of 3^2, 4^2 and 5^2

14 Add the prime numbers between 40 and 50.

Mathematical Reasoning

15 How many 4 cm cubes would fit inside this prism?

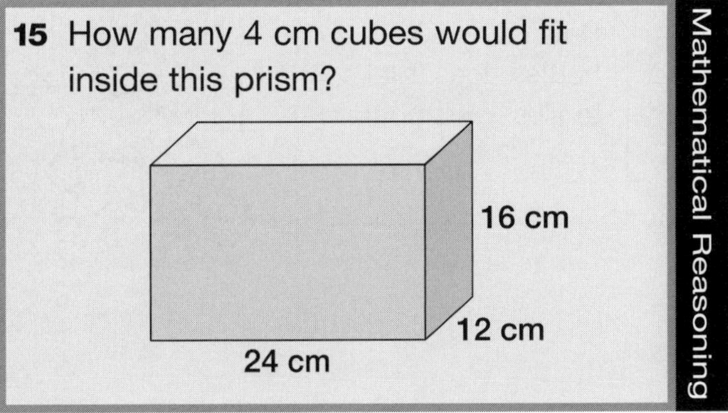

Measurement Metres, centimetres and millimetres

Estimate these measurements.

1	The length of your classroom	
2	The length of your bedroom	
3	The length of your pencil	
4	The length of a sultana	
5	The length of your shoe	
6	The distance from school to home	
7	The height of your teacher	
8	The width of a wedding ring	

Number and Algebra

SET 1 Basic

1 34 ÷ 6 = 5 r ☐

2 450 – 160

3 1 + 10 + 500 + 6000

4 (5 + 4) × 3

5 2.1 + 4.8

6 190 – 55

7 $\frac{1}{4}$ kg at $4.20 per kg

8 9 × 7

9 3.2 × 3

10 Quotient of 63 and 9

11 2.5 – 1.6

12 46 ÷ 5 = 9 r ☐

13 25 m + 38 m

14 Product of 8 and 12

15

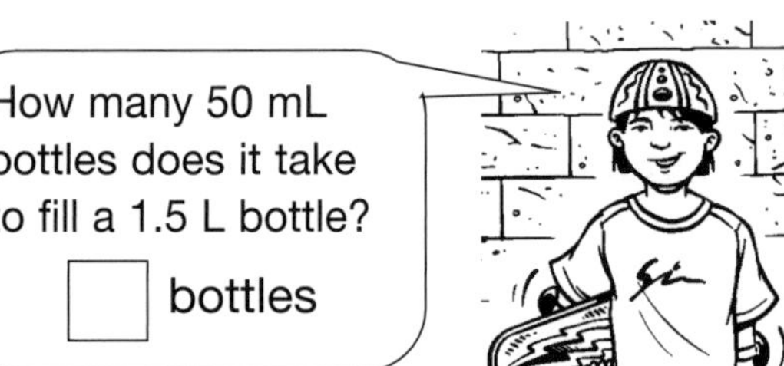

SET 2 Percentages

Calculate the saving on each item and the reduced price. (20% = $\frac{1}{5}$)

1

Joggers	
Saving	
Price	

2

Boots	
Saving	
Price	

3

High heel sandals	
Saving	
Price	

4

Thongs	
Saving	
Price	

Calculate these percentages.

5 25% of 80 children

6 50% of $48

7 10% of 70 litres

8 10% of 100 trees

9 75% of 40 goals

10 10% of $8000

Write a percentage for each fraction.

11 $\frac{1}{4}$ ☐

12 $\frac{4}{10}$ ☐

13 $\frac{1}{2}$ ☐

14 $\frac{1}{5}$ ☐

Measurement Area and perimeter

1 Design a square with an area of 16 cm^2.

2 Design a rectangle with a perimeter of 14 cm and an area of 12 cm^2.

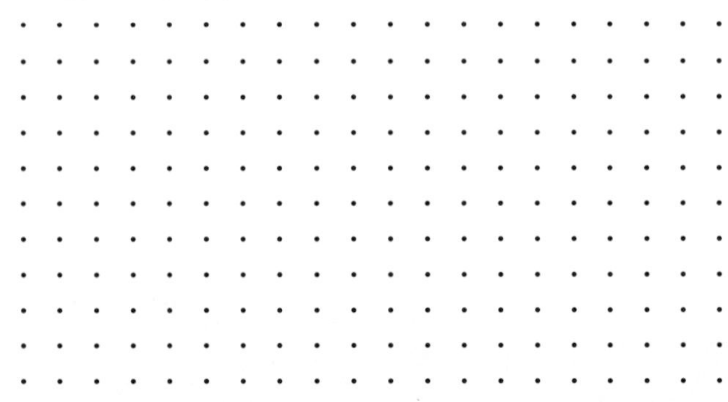

Number and Algebra

SET 3 Geometric patterns

Make a pattern of heptagons.

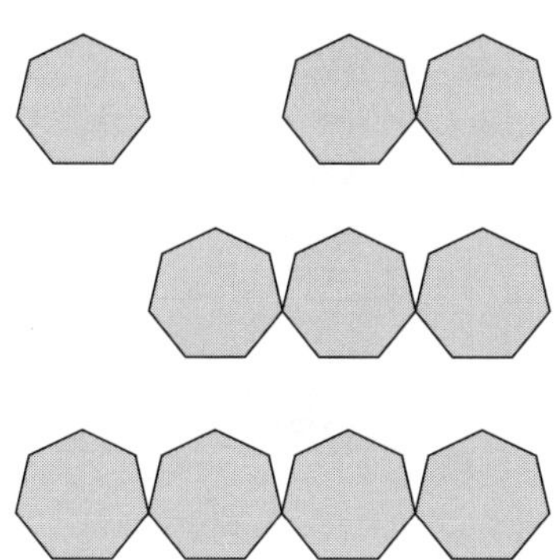

1 Complete and extend the table to record the number of sides needed to make the pattern of heptagons.

Heptagons	1	2	3	4	5	6	7
Sides							

2 Write a rule to describe the pattern.

____

3 How many sides would there be on 11 heptagons? ____________________

SET 4 Extension

1 Write 9:27 am in 24-hour time.

2 3.75 km = ☐ m

3 Write the prime numbers between 21 and 33.

4 Average $1.25, $3.75, $1.45

5 $\frac{\square}{4} = \frac{30}{40}$

6 How many degrees in 3 triangles?

7 Write fifty-five thousand, two hundred and seventeen in figures.

8 How many 1500 mL buckets are needed to fill a 24 L can?

9 What are the sides of a square if the area is 25 cm^2?

10 How many 175 mL bottles can be filled from 1.575 L?

11 Area of a rectangle with sides of 24 cm and 20 cm

12 $280, less 20%

13 What shape does this net make?

14 How much is 4.7 kg at $2 per kilogram?

15 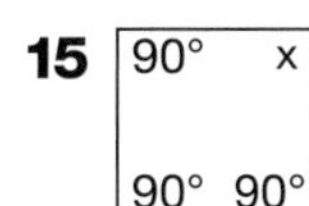

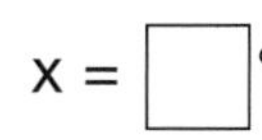

Statistics and Probability Graphs

1 In which sports is female participation higher than male participation?

2 In which three sports is female participation about equal with each other?

3 Do more girls play netball than boys play soccer?

4 Which three sports have the greatest number of players?

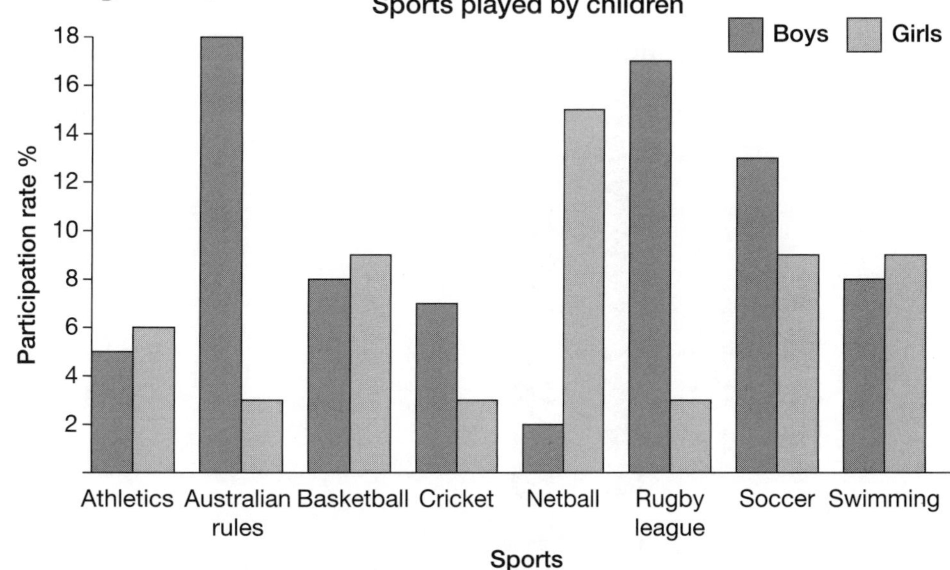

UNIT 23

Number and Algebra

SET 1 Basic

1 8 + 9 + 13
2 1000 ÷ 2
3 1000 – 600
4 1000 – 50
5 42 × 10
6 60 ☐ 2 = 30
7 8 ☐ 10 = 80
8 $3 \times 8 + 3^2$
9 Write the factors for 30.
10 Is 49 a prime number?
11 23 215, 23 230, 23 245, ☐
12 2 m and 97 cm = ☐ cm
13 5 hundreds + 2420
14 How many hours from 11 am to 3 pm?
15

SET 2 Multiplying by tens

1

×10	
7	
11	
26	
35	
20	

2

×10	
80	
90	
100	
127	
316	

3

×100	
8	
16	
23	
100	
118	

4

×1000	
6	
15	
50	
100	
225	

Mathematical Reasoning

Calculate the price of 100 items if you know how much 10 cost.

	Item	Cost of 10	Cost of 100
5	Football	$175	
6	Racquet	$238	
7	Bat	$334	

Statistics and Probability Dot plots

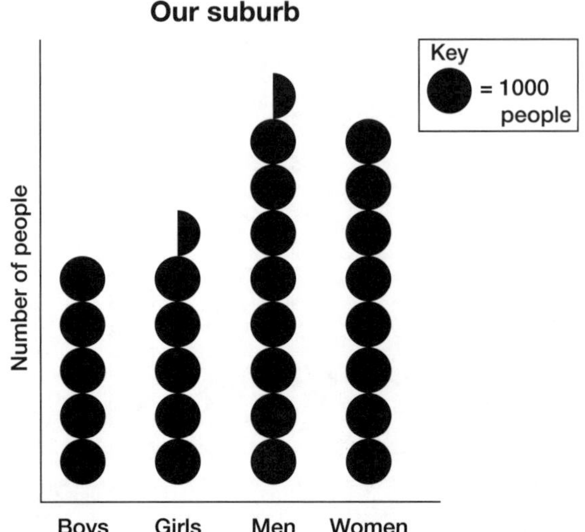

How many of each group lived in the suburb?

1 Boys? ______
2 Girls? ______
3 Men? ______
4 Women? ______
5 How many more men were there than boys? ______

Number and Algebra

SET 3 Rounding money

Round each price to the nearest dollar to estimate the cost.

1 3 kg at \$4.10 per kilogram.

2 5 kg at \$5.20 per kilogram.

3 6 kg at \$1.20 per kilogram.

4 8 kg at \$1.50 per kilogram.

5 12 kg at \$2.80 per kilogram.

6 10 kg at \$4.90 per kilogram.

7 5 kg at \$7.25 per kilogram.

8 8 kg at \$8.75 per kilogram.

9 9 kg at \$4.49 per kilogram.

Round the quantity bought to the nearest kilogram and the price to the nearest dollar to estimate the cost.

10 2.8 kilograms at \$5.05 per kilogram. 

11 5.25 kilograms at \$9.90 per kilogram. 

12 19.5 kilograms at \$9.99 per kilogram. 

SET 4 Extension

1 Write 10:15 pm in 24-hour time.

2 Order 3.3, 3.19, $3\frac{1}{5}$, $3\frac{1}{3}$

3 20% of \$1500

4 What fraction of 3 kg is 600 g?

5 $10^3 + 10^2 + 10 + 1$

6 How much is 6.7 kg at \$5 per kg?

7 Reduce 180 by 40%

8 Subtract the product of 9 and 4 from 9^2.

9 I was paid \$15 per hour and I worked from 8:45 am till 12:45 pm. How much did I earn?

10 48 minutes before 21:15

11 Difference between 27.6°C and 34.1°C

12 I went to bed at 10:30 pm and slept for $8\frac{1}{4}$ hours. When did I get up?

13 There are 12 rows of 108 seats. How many seats are there altogether?

14 Average 8.2, 9.4, 3.5 and 2.9

15 What is the perimeter of this trapezium?

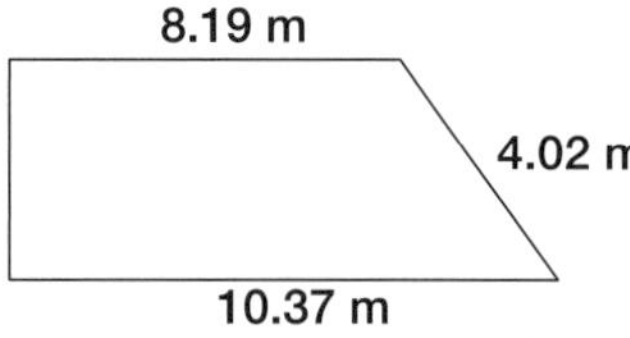

Statistics and Probability Pie charts

Dawn spends one hour each night at swimming training.

True or *false*?

1 About 30 minutes is spent on freestyle.

2 About 13 minutes is spent on backstroke.

3 About 7 minutes is spent on butterfly.

4 About 10 minutes is spent on breaststroke.

5 About half her time is spent on backstroke, breaststroke and butterfly.

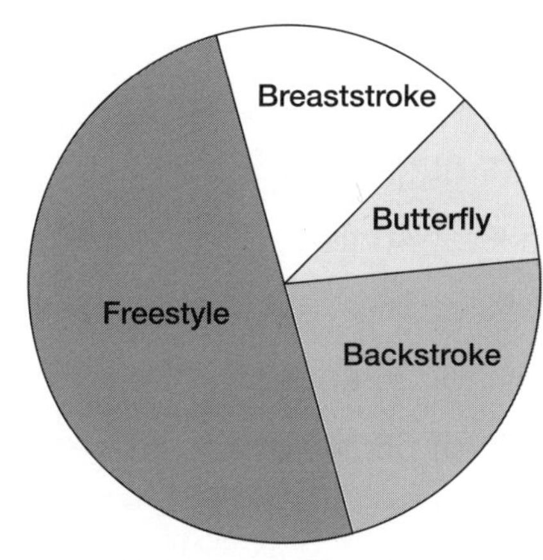

UNIT
24

Number and Algebra

SET 1 Basic

1 8 + 14 + 6

2 Is 17 a prime number?

3 18 ☐ 2 = 36

4 19 × 10

5 1900 − ☐ = 1000

6 64 ÷ 8

7 Tens in 5460

8 Factors of 27

9 \$8.15 × 100

10 85 × 100

11 Average of 6, 9 and 12

12 72 ÷ 9

13 Hundreds in 5289

14 Value of 8 in 85 962

15 150 cm × 3

16

The show starts at 10:15 and finishes at 11:30. How long is the show?

☐ minutes

SET 2 Multiplying decimals and money

1 13.2 × 4

2 22.4 × 5

3 20.7 × 6

4 14.32 × 5

5 \$10.08 × 7

6 \$25.86 × 8

7 213.7 × 3

8 224.07 × 6

9 \$243.82 × 5

10 Five children need 1.75 m of material each to make costumes for a play. How much material will they need altogether?

☐ m

11 A farmer has a large property which measures 7 km by 8.08 km. What is its area?

☐ km^2

Number and Algebra Finding percentages

1 10% of \$30

2 10% of 40 matches

3 20% of 30 pens

4 20% of 20 dogs

5 25% of 40 fish

6 25% of \$60

7 50% of 24 sheep

8 20% of 60 pencils

200 spectators watched a cricket match. Calculate the following numbers of people.

9 75% wore hats

10 90% wore sunglasses

11 60% had soft drinks

12 50% bought pies

13 30% cheered for the winners

14 75% drove to the match

Number and Algebra

SET 3 Number patterns

Follow the rule to complete the grids.

1 ■ = ▲ × 8

▲	7	6	5	4	3	2	1	0
■								

2 ★ = ◆ × 6

◆	3	6	9	12	15	18	21
★							

3 ▲ + 99 = ■

▲	11	12	13	14	15	16	17
■							

4 ▲ ÷ 2 = ■

▲	200	180	160	140	120	100	80
■							

5 Create a pattern and write a rule to explain it.

▲	1	2	3	4	5	6	7
■							

Mathematical Reasoning

SET 4 Extension

1 7.45 – 3.23

2 Which is not equivalent, $\frac{2}{5}$, 0.35, 0.4 or 40%?

3 Average 3.7, 4.5, 4.3, 3.5

4 4.5 + ☐ + 3.6 = 13.5

5 $\frac{8}{10} - \frac{71}{100}$

6 $1 - \frac{2}{100}$

7 How many sides on a dodecagon?

8 $\frac{35}{5} = \frac{\square}{10}$

9 How many 375 mL soft drink cans are needed to fill a 3 L bucket?

10 How many thousands in 236 107?

11 110°, x, 70°, 110° x = ☐°

Calculate the difference in hours and minutes between these times.

12 15:50 and 5:30 pm ______________

13 14:15 and 4:50 pm ______________

14 21:10 and 11:45 pm ______________

15 17:50 and 7:05 pm ______________

Mathematical Reasoning

Measurement Time

Draw the time each person finished work on the clock faces.

Ben started work at 9:00 am and finished $9\frac{1}{2}$ hours later.	Jessica started work at 8:15 am and finished $10\frac{1}{2}$ hours later.	Kelly started work at 7:45 am and finished 9 hours and 8 minutes later.	Stephen the night worker started at 9:30 pm and worked for 11 hours and 22 minutes.
1 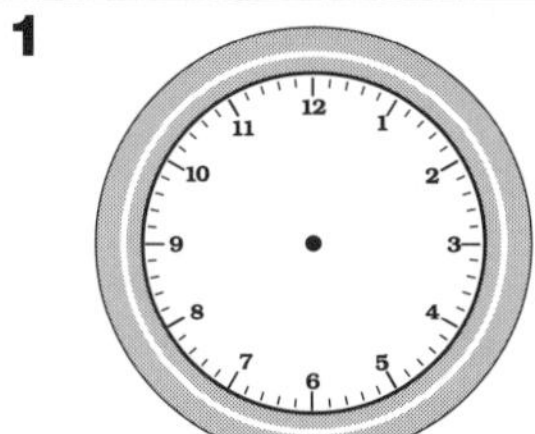	2 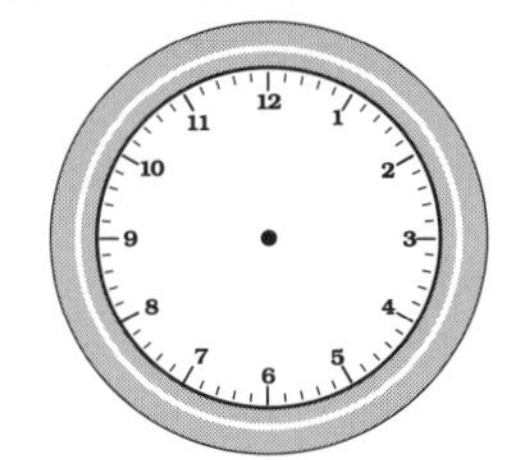	3 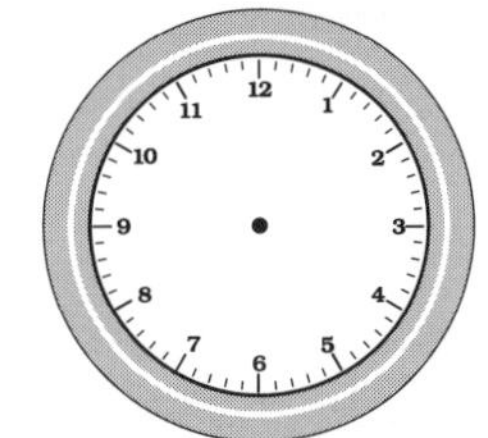	4 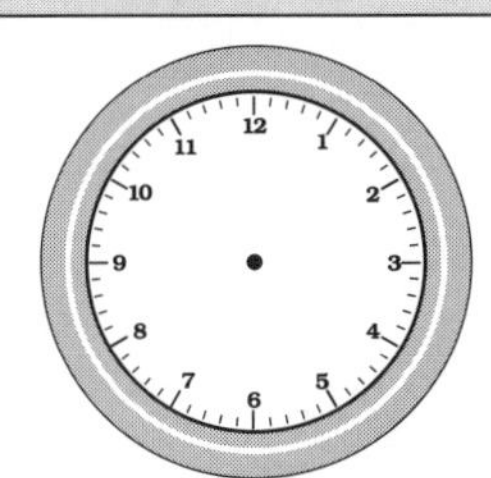

UNIT 25

Number and Algebra

SET 1 Basic

1 7 ☐ 23 = 30

2 7×9

3 $13c \times 100 = \$$ ☐

4 $\frac{2}{10} + \frac{5}{10}$

5 $5 + 25$

6 80 ☐ 50 = 30

7 63 ☐ 9 = 7

8 Halves in $1\frac{1}{2}$

9 0.3×2

10 $3 \times 7 + 3^2$

11 Is 36 a multiple of 5?

12 Is 35 a prime number?

13 235 306, 235 311, 235 316, ☐

14 How many minutes in $\frac{1}{2}$ hour?

15 How much change would I get from \$50 if I bought 8 chocolates at \$3 each?

\$ 

SET 2 Related denominators

Add and subtract the fractions with related denominators.

1 $\frac{1}{8} + \frac{1}{4}$

2 $\frac{3}{10} + \frac{2}{5}$

3 $\frac{1}{6} + \frac{2}{3}$

4 $\frac{5}{8} - \frac{1}{4}$

5 $\frac{9}{10} - \frac{1}{5}$

6 $\frac{7}{9} - \frac{1}{3}$

7 $\frac{9}{10} - \frac{3}{5}$

8 $\frac{7}{8} - \frac{1}{4}$

9 $\frac{5}{8} + \frac{3}{4}$

10 $\frac{9}{10} - \frac{1}{2}$

11 $\frac{1}{8} + \frac{1}{4} + \frac{1}{2} =$ ☐

Space Grid references

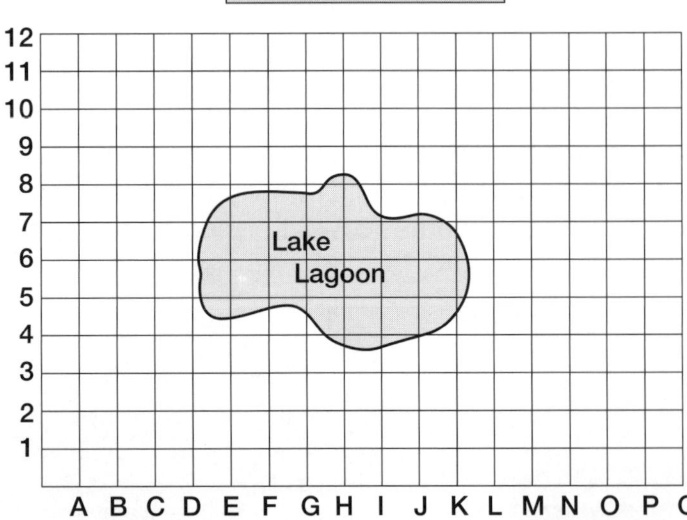

Put a dot on the map for each person's house.

1 Kelly A1

2 Ben A10

3 Taryn N10

4 Jim N3

5 Fred F3

6 Lauren P6

Use the scale to calculate the distance between:

7 Kelly's house and Ben's house.

8 Ben's house and Taryn's house.

9 Taryn's house and Jim's house.

Number and Algebra

SET 3 Decimal number patterns

Complete the decimal counting patterns.

1	0.20	0.25	0.30			
2	0.33	0.35	0.37			
3	0.45	0.60	0.75			
4	1.22	1.32	1.42			
5	4.05	4.45	4.85			
6	10.07	10.15	10.23			

7 Complete the function machine.

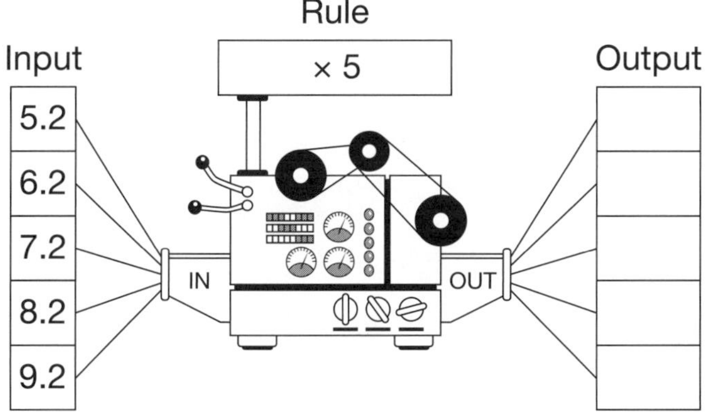

SET 4 Extension

1 $6^2 + 4^2$

2 $100 + 5 \times 4 + 79$

3 Add the difference between 180 and 130 to 70.

4 $30 + 20 \times 4 + 5$

5 $60 - 80 \div 5 + 6$

6 Which is the smallest, $\frac{7}{10}$, 75% or 0.71?

7 $2.6 \times 10 + 4$

8 $42.7 \div 7$

9 25% of $280

10 How much is 4500 g at $8.80 a kg?

11 1 m + 25 cm + 4 mm = ☐ mm

12 Round then estimate: 18.9×5.2.

13 How much interest is earned if $800 is invested at 10% p.a.?

14 Complete this sequence: 144, 121, ☐, ☐, ☐, 29

15 Factors of 42

16 A batsman hits 12 fours, 2 sixes and 47 singles. How many runs did he score?

17 John is $\frac{4}{5}$ of Tim's height. If Tim is 1.8 m, how much shorter is John?

Space Tessellations

Extend each tessellating pattern and name the 2D shapes.

1

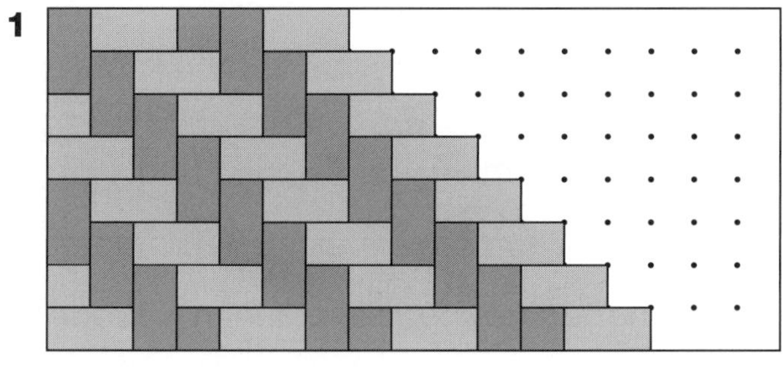

2D shape:

2

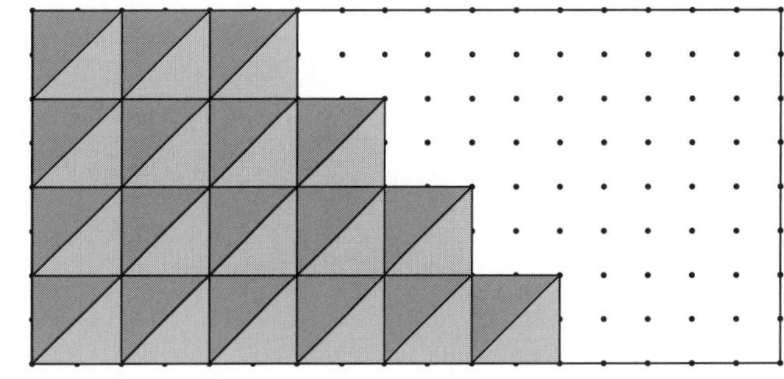

2D shape:

UNIT 26

Number and Algebra

SET 1 Basic

1 58c × 10

2 Factors of 45

3 0.4 × 6

4 $\frac{4}{10} + \frac{5}{10}$

5 Product of 9 and 12

6 How much is 250 g at $4.80 a kg?

7 3.4 × 4

8 Is 37 a multiple of 7?

9 Quotient of 81 and 9

10 4.2 × 10 + 18

11 $6^2 + 6$

12 59 ÷ 6

13 10:25 – 19 mins

14 Difference between 10^2 and 9^2

15

SET 2 Using factors to solve division

Supply the missing factors for the composite numbers below.

1
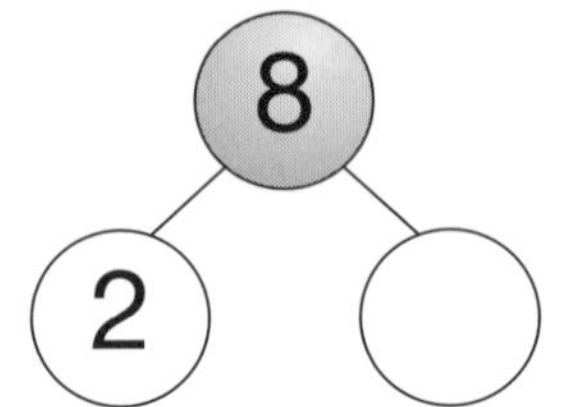

2
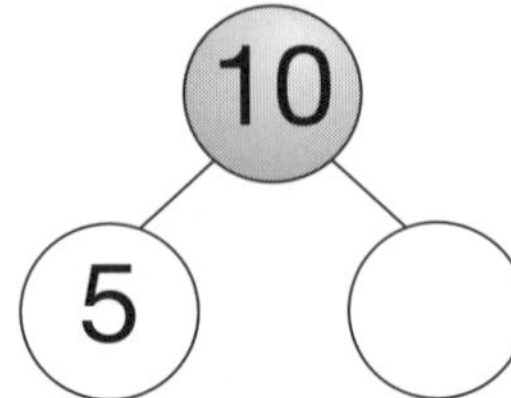

3
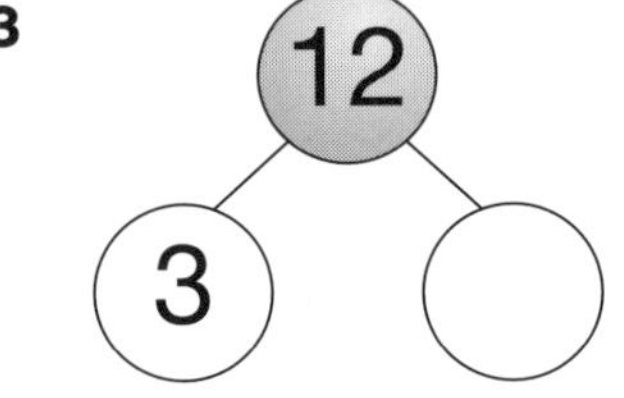

4
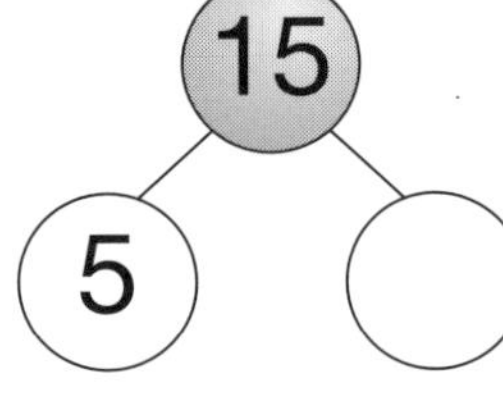

5
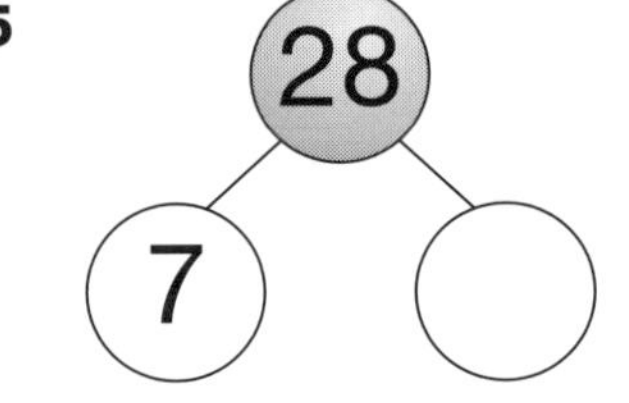

6
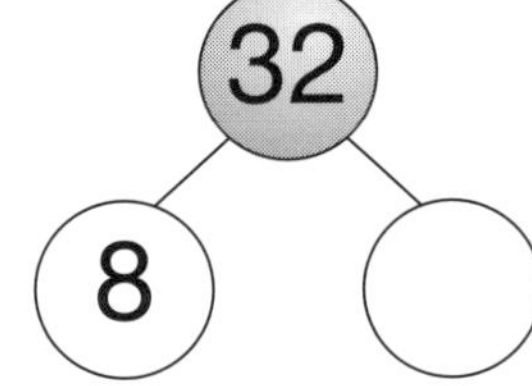

7
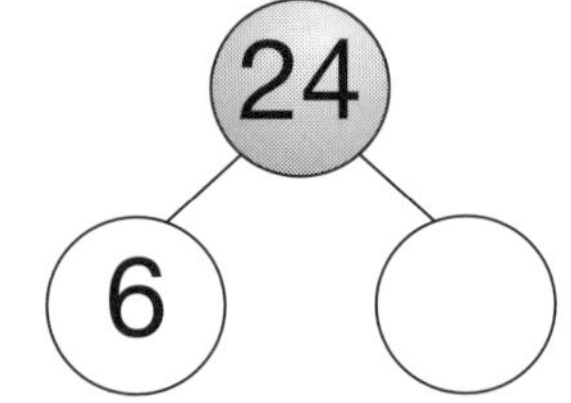

8
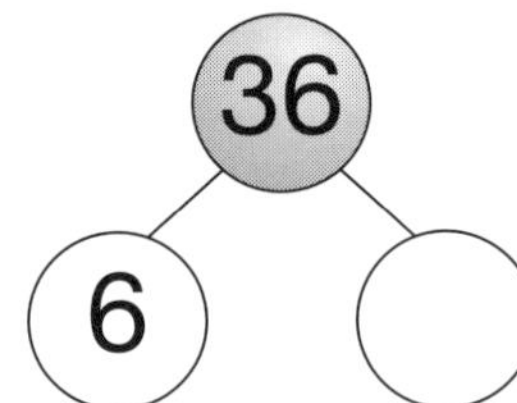

Measurement Measurement units

Convert these units into larger units.

1 450 mm = _____ cm

2 5000 mL = _____ L

3 9000 g = _____ kg

4 180 min = _____ hrs

5 1500 cm = _____ m

6 4000 mm = _____ m

7 3000 kg = _____ t

8 5000 m = _____ km

9 8000 mL = _____ L

10 240 min = _____ hrs

Convert these units into smaller units.

11 1.6 kg = _____ g

12 $\frac{1}{2}$ hr = _____ min

13 3.25 km = _____ m

14 4.05 m = _____ cm

15 $4\frac{1}{2}$m = _____ cm

16 2.5 hrs = _____ min

17 5.25 t = _____ kg

18 2 days = _____ hrs

19 3.25 m = _____ cm

20 0.565 kg = _____ g

21	How much did Ollie earn if he is paid $20 per hour and he worked for 90 minutes?	$

Number and Algebra

SET 3 Division of decimals

Divide these decimals.

1 $4\overline{)8.48}$

2 $3\overline{)6.96}$

3 $5\overline{)25.5}$

4 $5\overline{)45.5}$

5 $8\overline{)25.6}$

6 $4\overline{)9.24}$

7 $2\overline{)7.28}$

8 $6\overline{)7.38}$

9 $3\overline{)10.26}$

10 $4\overline{)17.64}$

11 Anita had a piece of ribbon that she wanted to cut into 5 equal pieces. If the ribbon is 26.55 m, how long would each piece be? 

12 Tara drew a rectangle that had an area of 36.9 cm^2. If one of the sides was 6 cm long what would the length of the other side be? 

SET 4 Extension

1 What is the difference in height between Kimberley (131 cm) and Jacqueline (191 cm)?

2 Perimeter of a rectangular field 65 m wide and 110 m long

3 Divide 240 m into 4 equal lots.

4 6 L of paint at $19.99 a litre

5 How many minutes in 12.7 hours?

6 Difference between 1 hr 14 min and 95 min

7 $\frac{3}{5} + \frac{2}{5} + \frac{7}{10} + \frac{1}{10}$

8 0.75 L $\times$ 100

9 60% of $150

10 $9.90 less 10%

11 How many litres of petrol could I buy with $30 if it cost $1.50 per litre?

12 Stuart has read $\frac{4}{5}$ of his book. What percentage has he left to read?

13 How many 275 mL jars can be filled from 2.2 L?

14 $\frac{8}{10}$ of 3 L

15 An intravenous drip releases 3 mL of fluid every 20 seconds. How much will it release in $1\frac{3}{4}$ hours? 

Statistics and Probability Chance

Draw a line to match each symbol shown on the spinner to a fraction, percentage or decimal that best describes the likelihood of that symbol being the winning symbol.

15%	30%	0.2	$\frac{1}{10}$	$\frac{1}{4}$

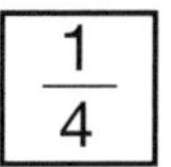

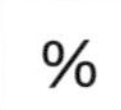

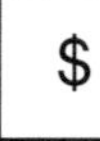

%	#	$	©	⊞

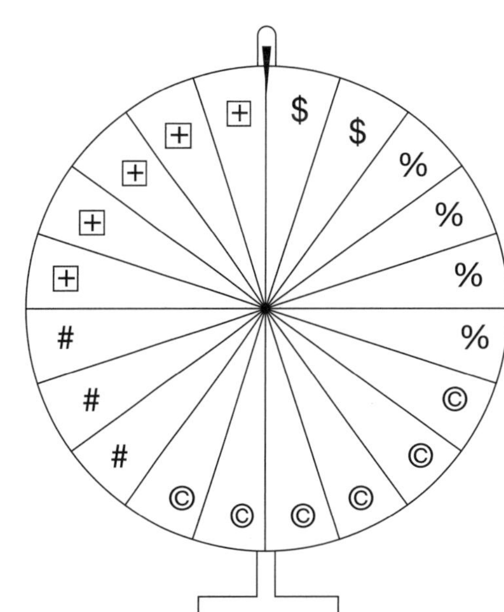

UNIT 27

Number and Algebra

SET 1 Basic

1 49 + 17
2 $6^2 + 18$
3 (19 – 11) × 7
4 25 – 6 + 1
5 Add 15 minutes to 10:55 am.
6 Subtract 7^2 from 100.
7 72c × 100
8 Factors of 21
9 Find the sum of odd numbers less than 10.
10 Find the sum of even numbers less than 10.
11 105 × 10
12 1000 ÷ 10
13 37° – 9°
14 $5.00 – $3.78
15

SET 2 Decimals × powers of ten

Multiply these decimals by powers of 10.

	Decimal	Multiply by	Answer
1	0.3	10	
2	0.7	10	
3	0.9	10	
4	0.27	10	
5	0.54	10	
6	0.34	100	
7	0.566	10	
8	0.876	100	
9	0.543	1000	
10	0.765	1000	
11	0.432	1000	
12	1.23	10	
13	1.25	100	
14	2.274	1000	

Space Angles

Calculate the size of each angle by subtracting the given angle from 180°.

1
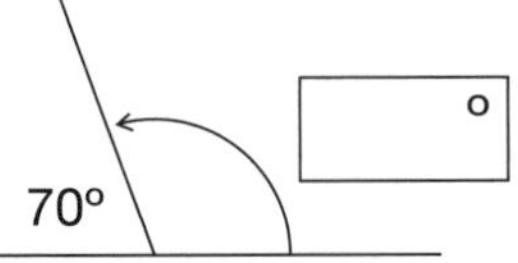

2
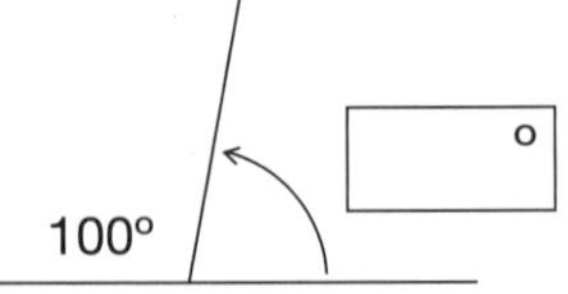

3
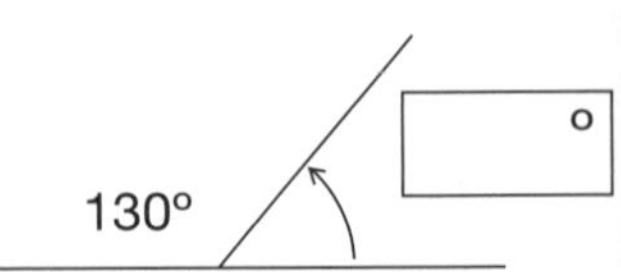

Calculate the size of each reflex angle by subtracting the given angle from 360°.

4
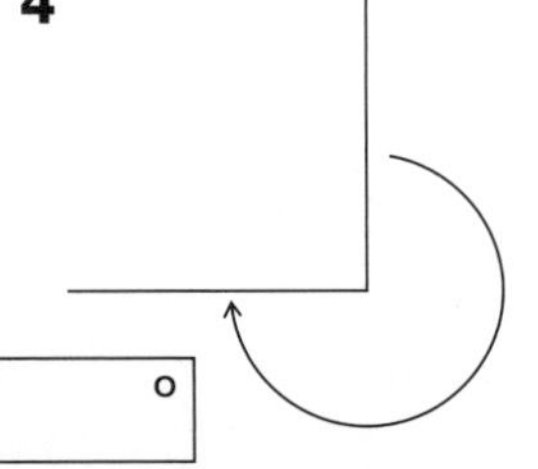

5
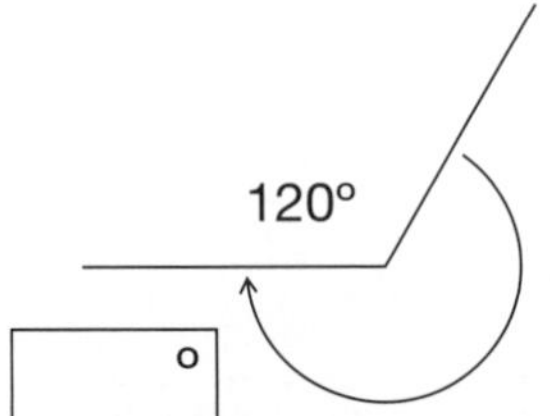

6
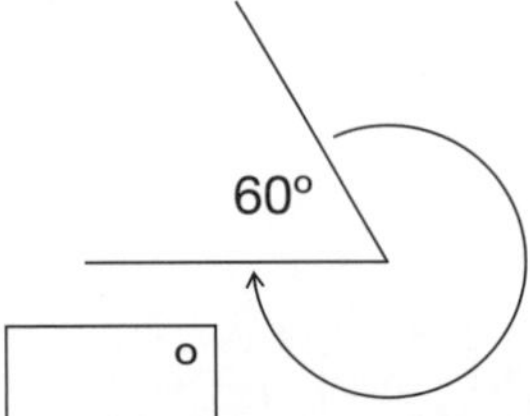

Number and Algebra

SET 3 Spreadsheets

1 Complete column D to show the balance after each transaction.

Willow's May spending

	A	B	C	D
1	Date	Item	Cost	Balance
2	1 May	opening	—	$1200.00
3	2 May	drink	$100.00	
4	8 May	bills	$75.00	
5	10 May	food	$125.00	
6	12 May	car	$40.50	
7	16 May	drink	$175.50	
8	20 May	food	$27.30	
9	30 May	bills	$99.70	
10	31 May	car	$40.00	

2 Does Willow have enough money left to buy a $600 TV set?

3 How much money had Willow spent by 10 May?

4 What was the total amount Willow spent on her car during May? 

$ ☐

5 What was the total amount Willow spent on food during May? 

$ ☐

SET 4 Extension

1 What is the perimeter of a quadrilateral with sides of 7.6 m, 3.4 m, 4.1 m and 5.2 m?

2 Which does not belong: 90%, $\frac{9}{10}$, $\frac{91}{100}$ or 0.9.

3 How many combinations of clothing can be made from 3 shirts and 3 pairs of pants?

4

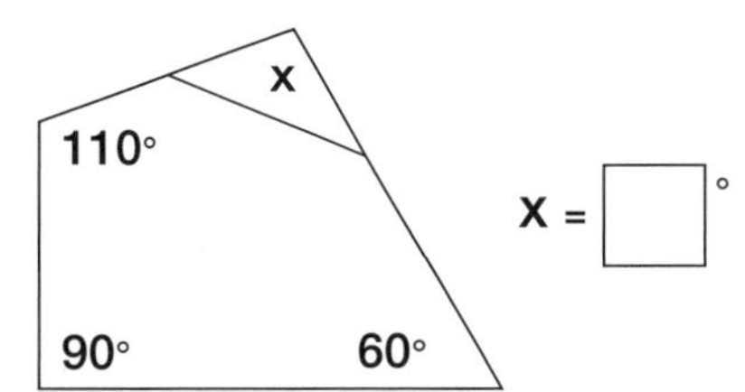

x = ☐°

5 Change $\frac{43}{5}$ to a mixed number.

6 Round 49 899 to the nearest 1000.

7 Dad's car covers 12 km per litre. If petrol costs $0.85 per litre, how much will a trip of 408 km cost? 

Mathematical Reasoning

8 Create a question that matches this working out.

168

7 × 24 × 60 = ______

______

______

______

Number and Algebra Graphing patterns

1 Complete the table to show the number of matches used to create the pattern of triangles.

Triangles	1	2	3	4	5	6	7	8
Matches	3	5						

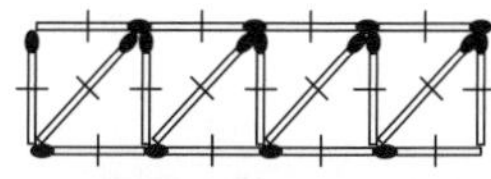

2 Plot the data in the table onto the graph.

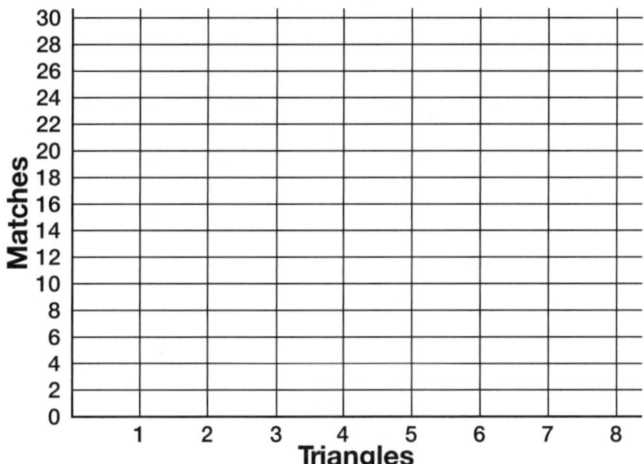

Number and Algebra

SET 1 Basic

1 25 ☐ 75 = 100

2 9 × 9

3 400 ☐ 250 = 150

4 18 + 24

5 80 ☐ 10 = 8

6 36 – 23

7 13 ☐ 3 = 39

8 \$17.45 = ☐ c

9 4526 + 3000

10 How much change from \$15 would I receive if I spent \$7.50?

11 How much are 5 books at \$1.40 each?

12 How many $\frac{1}{4}$s in 2?

13 How many tens in 150?

14 44 + ☐ = 5 × 10

15

SET 2 Estimation strategies

Tick the box to decide whether the statements are **Accurate** or **Inaccurate**.

	Statement	Accurate	Inaccurate
1	5 pens @ \$4.95 each would cost about \$25.		
2	My change from \$50 would be about \$30 after spending \$29.99.		
3	50 matchsticks would be enough to make 10 pentagon shapes.		
4	It would take 3 hours for a car to travel 90 km at 60 km/hr.		
5	If 6 movie tickets cost \$48, ten tickets would cost \$480.		
6	If a litre milk carton was half full, there would be 250 mL left.		

Statistics and Probability Frequency tables

Complete the frequency column on the table and transfer the data to the column graph.

Sports played by Year 6

Sport	Tally	Frequency
Soccer	卌 卌 ||	
Cricket	卌 |||	
Baseball	||||	
Netball	卌 卌 |	
Rugby	卌 |	
Tennis	|||	
Total		

Sports played by Year 6

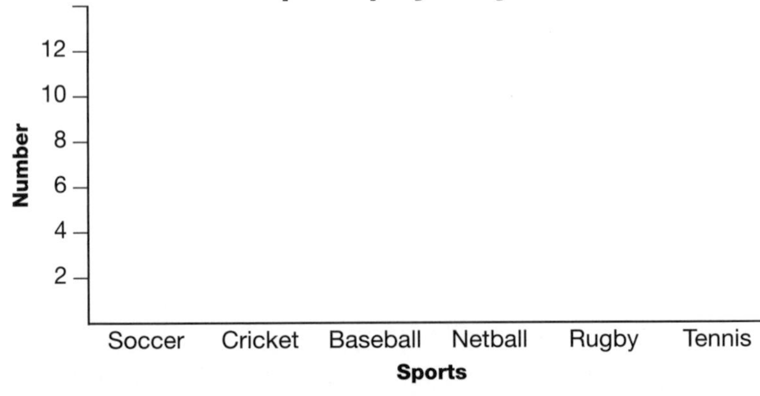

Number and Algebra

SET 3 Dividing decimals

1. $2\overline{)6.68}$
2. $7\overline{)49.7}$
3. $7\overline{)35.7}$
4. $10\overline{)85.50}$
5. $3\overline{)24.6}$
6. $4\overline{)3.24}$
7. $8\overline{)64.24}$
8. $7\overline{)35.14}$
9. $4\overline{)28.12}$
10. $3\overline{)2.16}$
11. $9\overline{)7.29}$
12. $6\overline{)55.26}$

13 Colour the label that gives the best value.

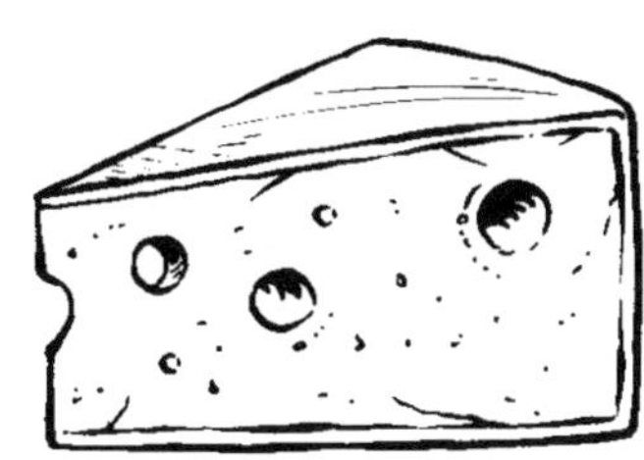

1 kg for $6.50

10 kg for $55.00

SET 4 Extension

1. Round 85 961 to the nearest hundred.
2. Value of 9 in 4.8907
3. 65% of 100 + 10% of 200
4. Twenty-eight minutes later than 9:56
5. 85% = [0.   ]
6. $17 + \frac{1}{2} \times 96 - 33$
7. 45% of $900
8. 64.48 ÷ 8
9. Quotient of 69 and 9
10. What is the area of a square with a perimeter of 20 cm?
11. 85 minutes after 2345
12. 900 m + 20 m + 3 km = [   ] km
13. How many degrees in 2 circles?
14. 90% of $250
15. Order 375%, 3.7, $3\frac{8}{10}$ and 3.79.
16. If 7 books cost $63.49, how much for 9?

Space The Cartesian plane

Plot these ordered pairs on the part of the Cartesian plane supplied.

The first one is done for you.

1. (5, 7)
2. (4, 6)
3. (3, 5)
4. (2, 4)
5. (1, 3)
6. (0, 2)
7. (–1, 1)
8. (–2, 0)
9. Draw a line to join the dots.

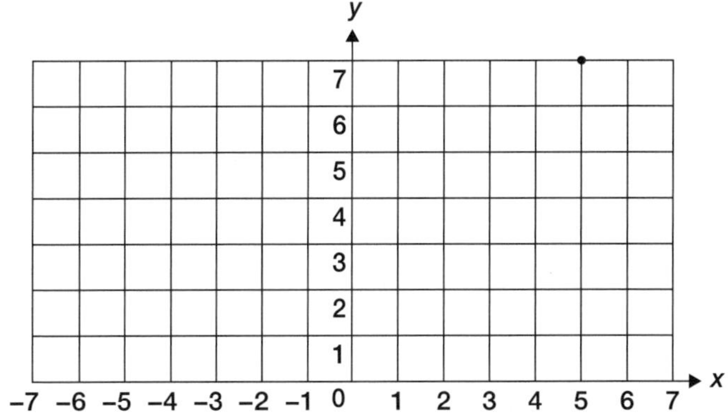

Number and Algebra

SET 1 Basic

1 Difference between 1000 and 20

2 Product of 8 and 20

3 $1.65 × 10

4 3.4 + 6.2

5 8 – 1.5

6 0.25 kg = ☐ g

7 How many hundreds in 5105?

8 How much is 2 L at $1.30 per litre?

9 Triple 25.

10 Divide 71 by 8.

11 $\frac{90}{100} - \frac{4}{10}$

12 98 + 150 + 2

13 10c × 1000

14 4 × ☐ = 1000

15

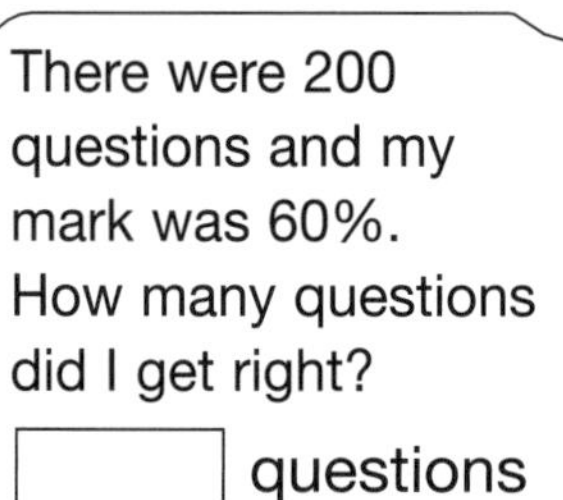

SET 2 Problem solving

Tick the number sentence that would solve the problem.

1 30 children are going on an excursion to the aquarium. Entry tickets are $5 per person. How much does the teacher need to collect if she also needs $300 to hire a bus?

	($5 × 30) + $300
	($5 + 300) × 30

2 How much money did Year 6 raise if 80 children paid $3.50 to attend the disco and another $350 was donated by a sponsor?

	$3.50 + 80 + $350
	(3.50 × 80) + $350

3 How many millilitres of juice are left in the 5 litre container if 2 litres was poured into a jug and ten 200 mL cups were filled?

	(5000 mL – 2000 mL) – 200 mL × 10
	(5000 mL – 2000 mL) – 200 mL

4 Noah's step is 75 cm and Joni's step is 65 cm. How much further ahead will Noah be if they each take 20 steps?

	(75 cm + 65 cm) × 20
	(75 cm – 65 cm) × 20

5 Mali made 6 hexagons and 2 octagons using matchsticks for each side. How many match sticks did she need?

	(6 × 6) + (2 × 8)
	(6 × 8) + 6 × 2

Space Angles

Record the missing angle on each set of complementary and supplementary angles.

1

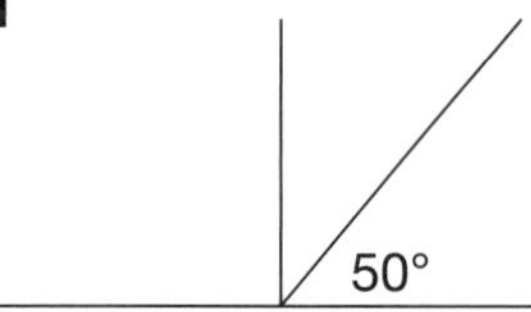

2

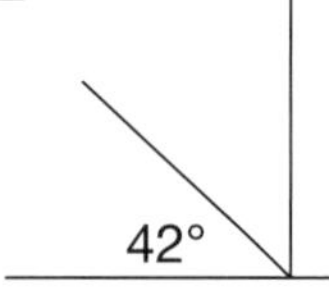

3

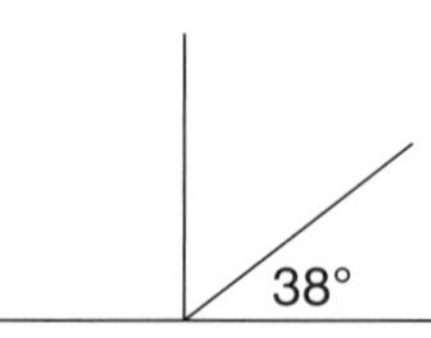

4

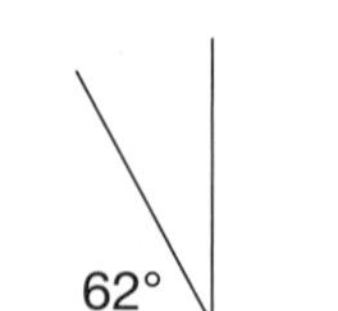

5

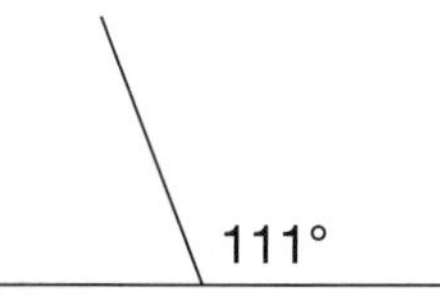

6

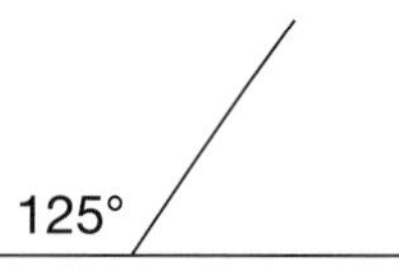

7

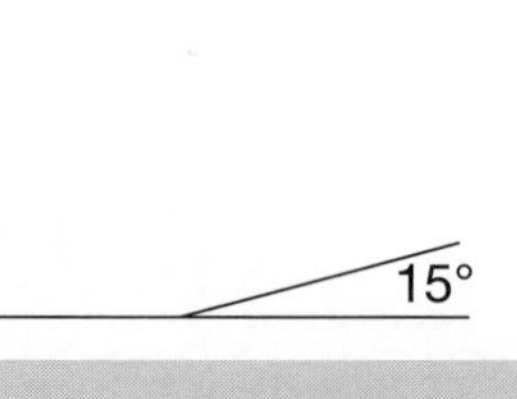

8

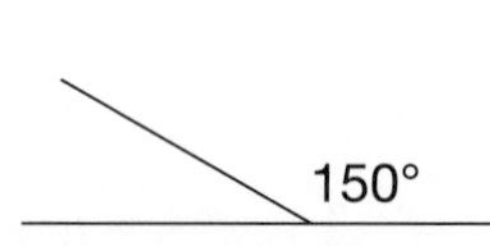

Number and Algebra

SET 3 Solving equations

Solve the equations and problems.

1 45 ÷ 9 × 6 =

2 7 + 9 × 7 =

3 36 – 49 ÷ 7 =

4 (47 – 20) ÷ 9 =

5 (47 + 19) × 3 – 40 =

6 7 × 9 + 5 =

7 7 × (9 + 5) =

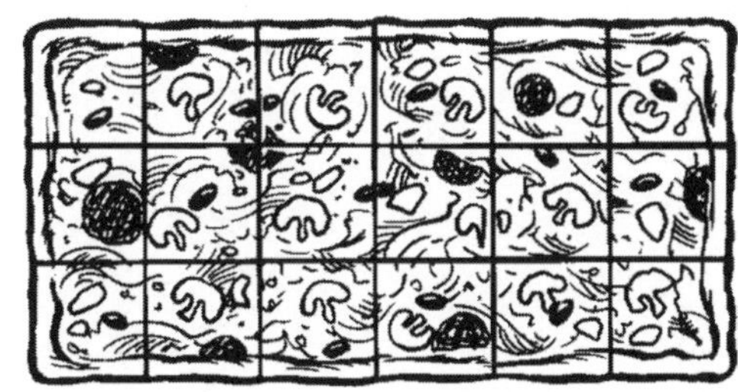

8 If Tom ate $\frac{3}{18}$, Jerome $\frac{5}{18}$ and Sally $\frac{4}{18}$, how much would be left?

Mathematical Reasoning

9 How could an 18 piece pizza be shared between two people so that one person receives twice as much as the other?

SET 4 Extension

1 11.11, 11.31, 11.51, ☐

2 How many 1.35 m lengths can be cut from 6.75 m?

3 Write 9% as a decimal.

4 $\frac{15}{42} = \frac{\square}{14}$

5 Name this shape:

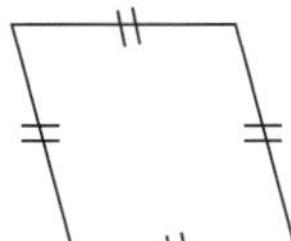

6 3.21 m = ☐ cm

7 How many edges has a cube?

8 Perimeter of a square with an area of 64 cm^2

9 x = ☐ °

x 60°

Mathematical Reasoning

Australia's population is 27 000 000

Use a calculator to work out how many people belong to each age group.

	Age group	Number
10	21% are under 15	
11	67% are aged 15–64	
12	12% are 65 and over	

Statistics and Probability Chance predictions

Matt dropped his bag of 24 marbles and 6 rolled out. Use the marbles that rolled out to predict how many marbles of each colour might be in the bag.

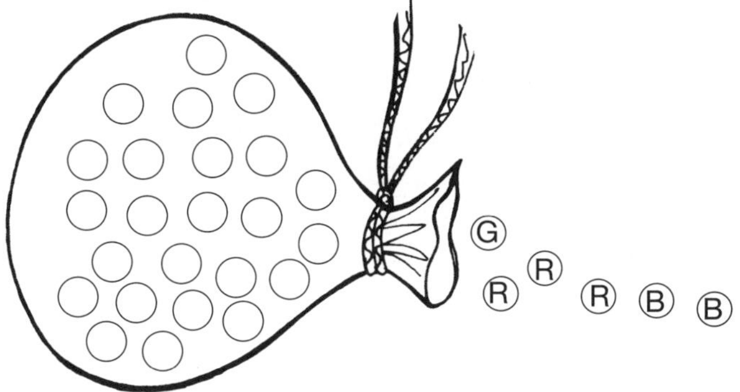

1 Red ☐

2 Blue ☐

3 Green ☐

UNIT 30

Number and Algebra

SET 1 Basic

1 How many mm in 1 m?

2 91×10

3 $4^2 + 5^2$

4 $36 = \square^2$

5 $700 - 350$

6 $12 \square 5 = 60$

7 $36 \square 14 = 50$

8 $21 \div 5$

9 Divide 60 by 5.

10 $6 \times 8 + 12$

11 Difference between 270 and 50

12 Add 37 and 137.

13 $69 + 30 + 1$

14 $\frac{1}{3}$ of 15

15

Melissa poured 2 L and 350 mL into a jug. How many mL were there altogether?

☐ mL

SET 2 Positive and negative numbers

Solve the number sentences.

1 $2 - 5 =$

2 $3 - 6 =$

3 $7 - 15 =$

4 $1 - 9 =$

5 $-2 + 5 =$

6 $-4 + 6 =$

7 $-5 + 7 =$

8 $-8 + 9 =$

9 $2 + 3 - 7 =$

10 $-7 + 3 + 5 =$

11 $-9 + 4 =$

12 $-10 + 7 =$

Statistics and Probability Graphs

Use the graph to decide whether these statements are true or false.

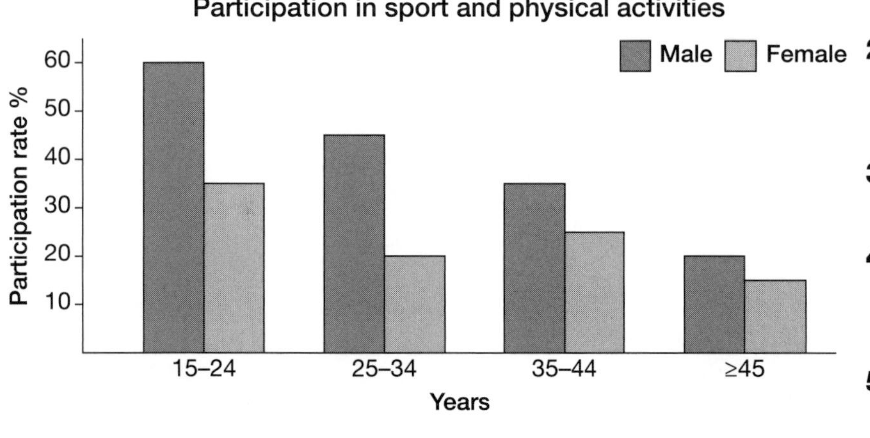

		T or F
1	65% of males aged 15–24 participated in sports.	
2	35% of females aged 15–24 participated in sports.	
3	45% of males aged 25–34 participated in sports.	
4	25% of females aged 25–34 participated in sporting activities.	
5	Females aged ≥45 had the lowest participation rate.	

Number and Algebra

SET 3 Prime factors

Complete the factor trees.

1

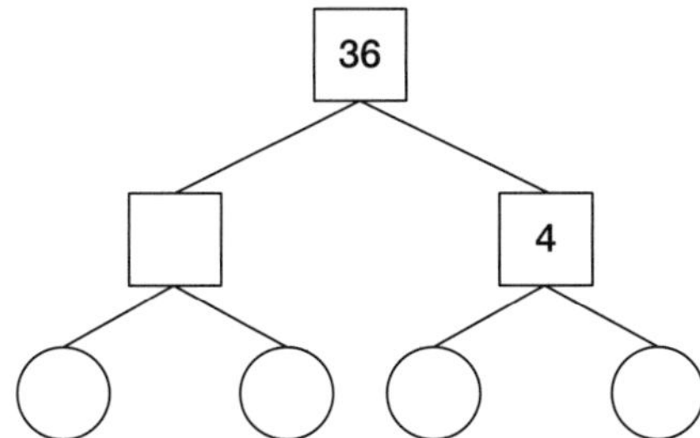

2

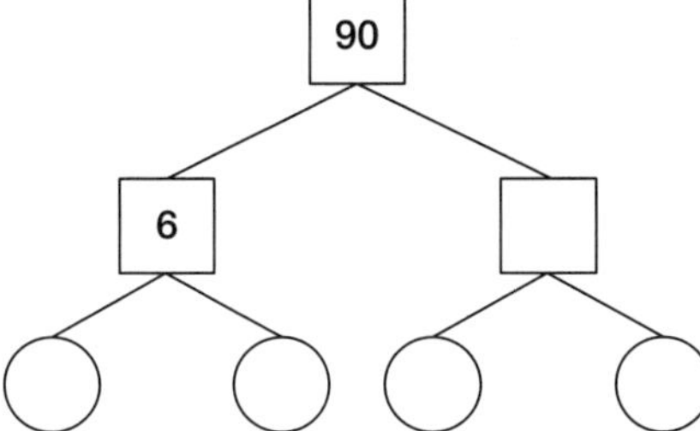

3

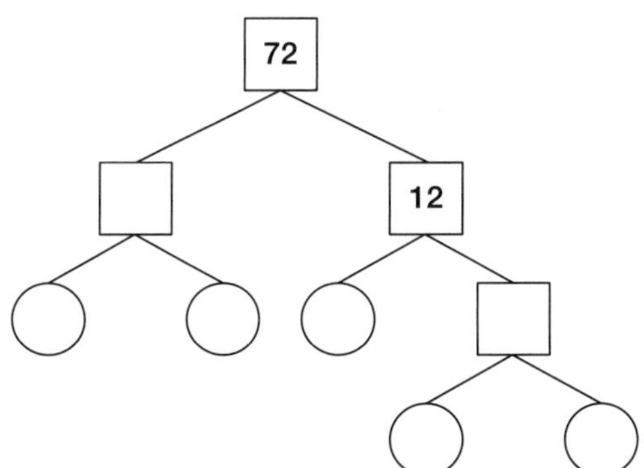

SET 4 Extension

1. 5 m + 85 cm = ☐ cm
2. How much is 3 kg at \$48 for 6 kg?
3. 74% = ☐ hundredths
4. 25% of \$2600
5. Round off and estimate 69.8 × 8.9.
6. 3 × 60 ÷ 9 + 7
7. What fraction of 250 is 75?
8. 56.16 ÷ 8
9. 2 m + 32 cm + 8 mm = ☐ mm
10. Quotient of 774 and 8
11. How many minutes between 9:55 am and 2:24 pm?
12. How much is 750 g at \$3.60 a kg?
13. What is the volume of a swimming pool with dimensions 9 m, 5 m and 2 m?
14. Round 1 987 621 to the nearest million.
15. 800 m + 100 m + 2 km = ☐ km
16. Mia won \$350 in Lotto. If she kept $\frac{3}{5}$ of it and gave the rest away, how much did she keep? \$ ☐

Space Vertically opposite angles

Calculate the size of the vertically opposite angles.

1

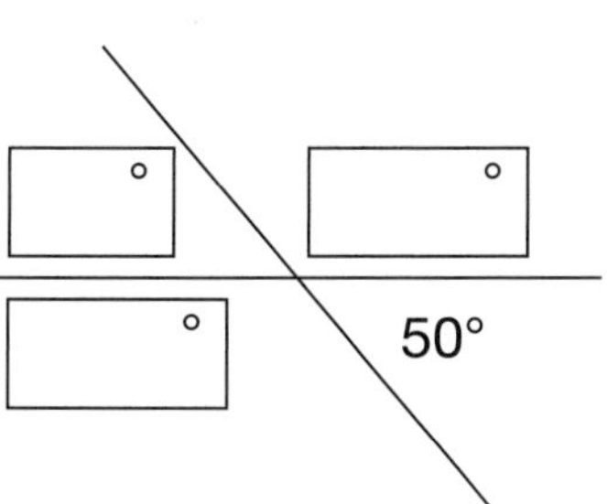

2

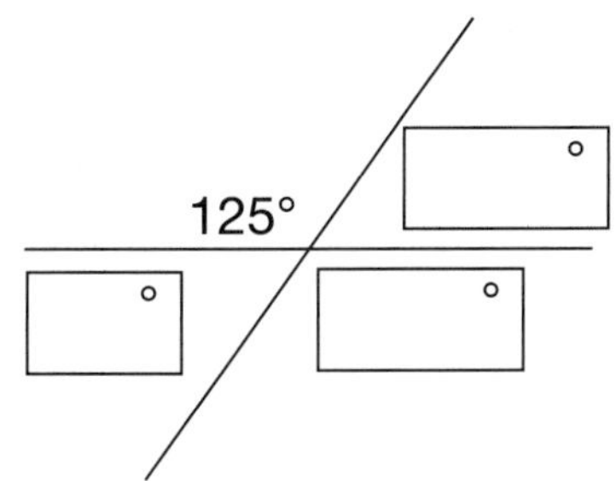

3

☐° ☐° ☐° 70°

Number and Algebra

SET 1 Basic

1 350 + 60

2 800 ☐ 2 = 400

3 Product of 6 and 10

4 Sum of 30 and 170

5 3×55c

6 Divide 43 by 6.

7 3 thousands + 17 231

8 Value of 9 in 923 231

9 Are 42 and 37 factors of 9?

10 List primes between 20 and 30.

11 Write the factors of 21.

12 $\frac{23}{100} = 0.$ ☐

13 $1 - \frac{3}{4}$

14 \$37.45 ☐ c

15

SET 2 Negative numbers

Find the difference in temperature between:

1 −3°C and 8°C _______°C

2 −2°C and 7°C _______°C

3 −5°C and 10°C _______°C

4 −10°C and 11°C _______°C

5 −5°C and 20°C _______°C

6 −7°C and 18°C _______°C

7 −8°C and 15°C _______°C

8 −7°C and 14°C _______°C

9 −1°C and 20°C _______°C

10 −9°C and 13°C _______°C

Solve the equation.

11 $6 - 5 + 2 - 8 + 7 - 2 =$

12 $0 - 6 + 2 + 6 - 9 + 3 =$

13 $-9 + 6 + 6 + 2 - 10 + 2 =$

14 $-6 + 3 + 7 - 5 + 3 - 8 =$

15 $-21 + 10 + 20 - 40 + 3 =$

16 $-37 + 20 + 33 - 30 + 5 =$

Space Tessellating shapes

Continue the tessellating pattern of irregular hexagons.

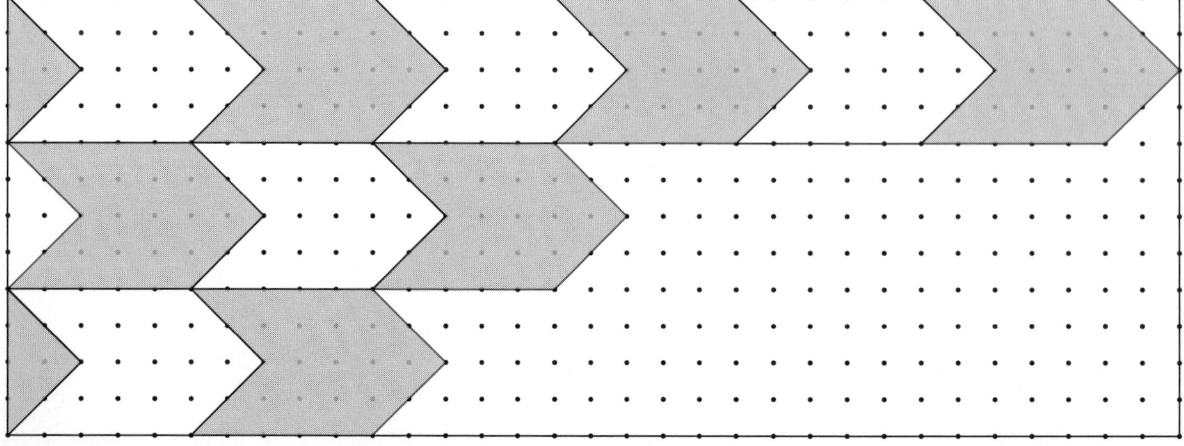

Number and Algebra

SET 3 Number patterns

Follow the rules to find the output of the machines.

1

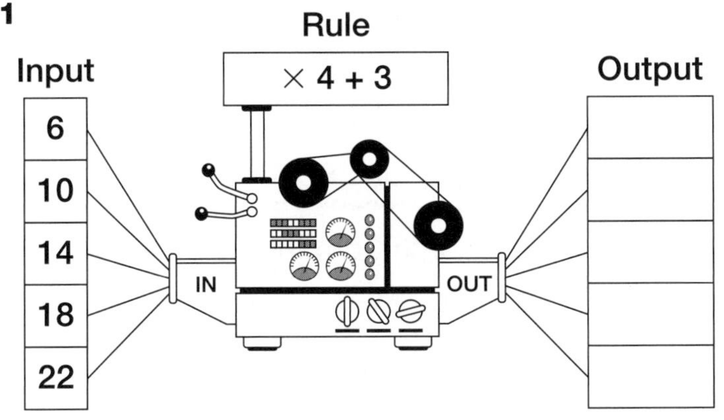

2

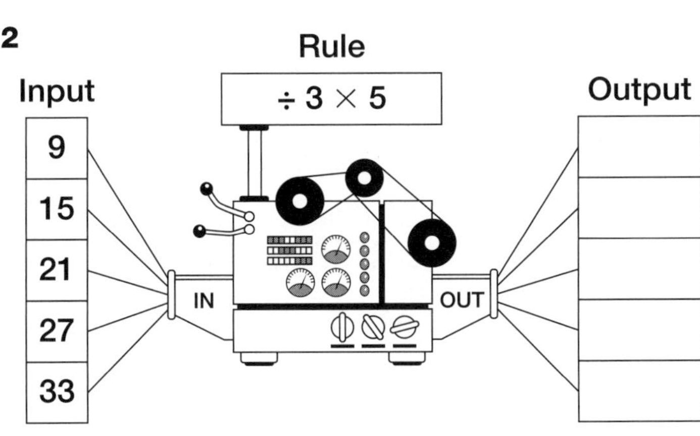

SET 4 Extension

1 5×10^2 + 18 tens

2 1, 2, 3, 5, 8, 13, ______

3 Average 11.4, 12.3, 9.3, 3.4

4 500 × 341 × 2

5 3.5 ha ÷ 7

6 9.5 ÷ 5 × ☐ = 11.4

7 Area of a square which has a 28 cm perimeter

8 $\frac{76}{100} - \frac{5}{10}$

9 Centimetres in 1 km

10 How many minutes are there from 9:48 pm to 1:05 am?

11 $800 less 40%

12 2 m = 1500 mm + ☐ cm

13 What time is it $\frac{3}{4}$ of an hour before ten to five?

14 What is the radius of a circle which is 186 mm wide?

15 4.5 km minus $\frac{8}{10}$ of a kilometre

16 How many cubes measuring 3 cm can fit inside a box measuring 9 cm by 6 cm by 12 cm?

Measurement Capacity units

1000 millilitres = 1 litre   1000 litres = 1 kilolitre   1 000 000 litres = 1 megalitre

Convert the measurement to another unit.

1 2000 mL = __________ L

2 5000 mL = __________ L

3 7 L = __________ mL

4 $3\frac{1}{2}$ L = __________ mL

5 $7\frac{1}{4}$ L = __________ mL

6 2000 L = __________ kL

7 8000 L = __________ kL

8 5 kL = __________ L

9 $2\frac{1}{2}$ kL = __________ L

10 4 ML = __________ L

Number and Algebra

SET 1 Basic

1 9^2

2 76c + 94c

3 80 × 5

4 Product of 9 and 3

5 40 ÷ 6

6 $28 × 100

7 $39 ÷ 3

8 0.65 = ☐ %

9 4 × ☐ = 448

10 7.7 + 2.13

11 0.8 + 0.8 + 0.8 + 0.8

12 Difference between 90 and 16

13 Product of 9 and 12

14 Value of 6 in 27.615

15

How many minutes are there from 9:15 am to 11:07 am?

☐ minutes

SET 2 Multiplication problems

Surf Supplies Incorporated 1998

Surfboard $399

Wetsuit $250

Fins $180

Rash shirt $79

Zinc $2.50

Wax $3.80

Calculate the income from the business last month if these sales were made.

	Item	Quantity	Cost per unit	Income
1	Board	38		
2	Wetsuit	29		
3	Fins	40		
4	Shirt	63		
5	Zinc	99		
6	Wax	95		

Calculate the total cost of these items.

7 2 boards and a shirt ________

8 fins, wax and zinc ________

9 wetsuit, fins and board ________

10 5 shirts and 3 zincs ________

Space The Cartesian plane

Write the ordered pairs for each letter.

1 A = (,)

2 B = (,)

3 C = (,)

4 D = (,)

5 E = (,)

6 F = (,)

7 G = (,)

8 H = (,)

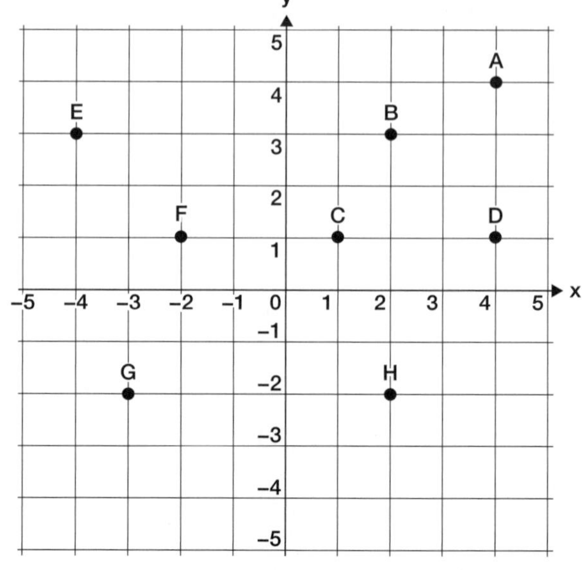

Number and Algebra

SET 3 Balance

Complete the number sentences by supplying the missing number.

1 6 + ☐ = 22 – 7
2 8 × ☐ = 9.6
3 36 ÷ ☐ = 54 ÷ 9
4 9 × ☐ = 13.5
5 ☐ × 4 = 75 – 55
6 13 × ☐ = 49 – 10
7 17 + ☐ = 96 – 66
8 7 × ☐ = 3.5
9 ☐ ÷ 7 = 81 – 75
10 6.4 ÷ ☐ = 0.8

Mathematical Reasoning

Supply the missing operation signs.

11 8 ☐ 5 ☐ 2 = 42
12 (8 ☐ 5) ☐ 2 = 26
13 (8 ☐ 5) ☐ 2 = 6
14 8 ☐ (5 ☐ 2) = 24

SET 4 Extension

1 $\frac{3}{8}$ of 96 = $\frac{3}{6}$ of 72 *True* or *false*?
2 Round 35 244 to the nearest 10.
3 How many 120 g masses would be needed to balance a 2.4 kg mass?
4 Average of 3.5, 2.5, 3.2 and 2.8
5 Round 25 369 to the nearest 1000.
6 What is the value of 6 in 6937.24?
7 Is body temperature about 5°C, 0°C, 100°C or 37°C?
8 Write five million, three hundred and two thousand in figures.
9 Area of a square with a 15 cm base
10 Estimate the answer for 39 987 × 3.
11 How much is 4.6 kg at $9 per 500 g?
12 75% of $300
13 4.35 tonnes = ☐ kilograms
14 How many seconds in 1.3 hours?
15 Round 450.97 to the nearest 100.
16 $\frac{1}{4}$ of 28 × $\frac{6}{10}$ of 130
17 How many sides are on 7 decagons?
18 What is the perimeter of a decagon with 17.2 cm sides?

Statistics and Probability Interpreting data

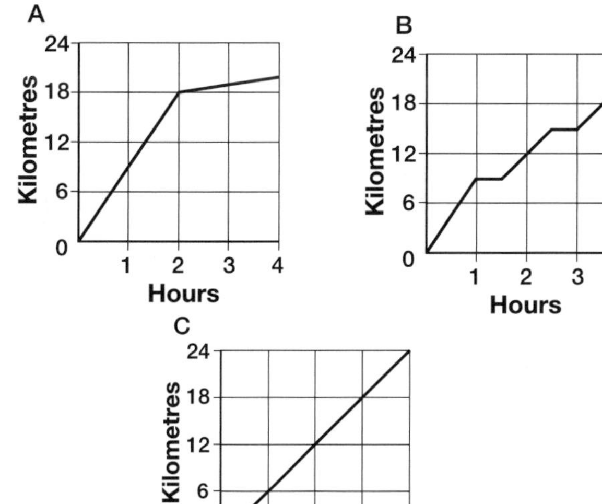

Study the graphs which show how three people performed in a 24 km fun run. Answer *true* or *false*.

1 After 1 hour A was at the 9 km mark. ☐
2 After 1 hour C had travelled 6 km. ☐
3 B had a 30 minute rest after 1 hour. ☐
4 A averaged 6 km/h for the first 2 hours. ☐
5 B averaged less than 6 km/h for the whole race. ☐
6 Slow and steady wins the race. ☐

Number and Algebra

SET 1 Basic

1 31 ÷ 6

2 2000 mL = ☐ L

3 34 × 100

4 9 ☐ 3 = 27

5 70 ☐ 80 = 150

6 8 × 8 + 6

7 60 – 45

8 Divide 15 by 4.

9 How many 100s in 3754?

10 What is the value of 5 in 8754?

11 $1\frac{1}{2}$ kg = ☐ g

12 I had $50 but spent $27. How much do I have left?

13 37 426 + 300

14 How much in $1\frac{1}{2}$ kg of meat at $8 per kg?

15 How much are 5 pots at $9 each?

16 How many prime numbers are there between 6 and 20?

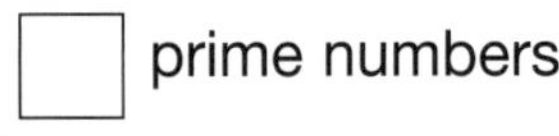

SET 2 Decimals × powers of ten

You may need to use your calculator for some of these questions.

	×	10	100	1000
1	0.231			
2	4.38			
3	0.643			
4	1.87			
5	15.6			

	÷	10	100	1000
6	3568			
7	4295			
8	235.2			
9	68.9			
10	19.5			

Write *true* or *false*.

11 3.68 × 10 = 36.8 ________

12 0.183 × 100 = 18.3 ________

13 23.6 ÷ 100 = 0.236 ________

14 18.5 ÷ 1000 = 0.0185 ________

15 2.81 ÷ 10 = 28.1 ________

16 0.392 × 1000 = 392 ________

Measurement Square and cubic metres

1 Calculate the area of the floor of the factory.

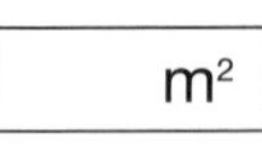

2 Calculate the volume of the factory.

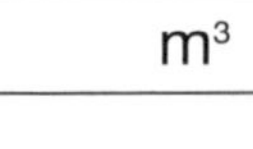

3 Calculate the cost of building the factory at $300 per m³. ☐ $³

Number and Algebra

SET 3 Decimal/fraction number patterns

Complete the sequences.

1	0.4	0.8	1.2			
2	4	4.16	4.32			
3	8	8.11	8.22			
4	4	4.23	4.46			
5	5	5.17	5.34			
6	6	6.03	6.06			
7	3	$3\frac{2}{3}$	$4\frac{1}{3}$			
8	4	$4\frac{3}{4}$	$5\frac{1}{2}$			
9	5	$5\frac{2}{5}$	$5\frac{4}{5}$			
10	6	$6\frac{3}{5}$	$7\frac{1}{5}$			

SET 4 Extension

1 8.75 tonnes = ☐ kg

2 How much is 14.25 kg of potatoes at $3 per kilogram?

3 Average 125, 275, 100 and 120.

4 All quadrilaterals are polygons. True or false?

5 How many combinations can I make with 4 shirts and 3 hats?

6 What percentage of $120 is $24?

7 x, 30°, 50° x = ☐°

8 John travelled for 6 hours and covered 522 km. What was his average speed?

9 18 m, less 180 cm

10 (17.86 + 2.14) × 6

11 3.26 × 100

12 Order these decimals: 0.1, 1.0, 0.11, 11.1

13 How much money did I begin with if I spent 75% of my money and had $50 left over?

____

Mathematical Reasoning

Statistics and Probability Two-way tables

Complete the totals for each column on the cricket scoresheet to show how many overs the bowlers have bowled, how many wickets they have taken and how many runs the opposition batters have hit.

Bowler	Overs	Wickets	Runs
Murray	8	2	16
Lane	7	0	34
Vanda	9	4	28
Simms	6	1	22
Wallace	5	3	20
Total			

1 Which bowler conceded the most runs?

2 Who was the most successful wicket-taker?

3 Did the bowler who bowled the most overs take the most wickets?

4 Did the bowler who bowled the least overs concede the least number of runs?

5 Do you agree that on average the team takes one wicket every 12 runs?

UNIT 34

Number and Algebra

SET 1 Basic

1 350 + 150

2 500 ☐ 100 = 5

3 360 – 90

4 40 ☐ 170 = 210

5 Product of 10 and 60

6 How many days in June, July and August?

7 $\frac{3}{100} = 0.3$ True or false?

8 76 × 1000

9 Tenths in $2\frac{1}{2}$

10 $8^2 + 14$

11 $30 - 5 = 5^2$ True or false?

12 Double 280.

13 Triple 20.

14 36 – 7 = (6 × 5) True or false?

15 If 4 balls cost $2.40, how much would 3 balls cost?

SET 2 Fraction and decimal remainders

Solve these divisions writing your remainder as a fraction.

1 $3\overline{)16}$ = $5\frac{1}{3}$

2 $4\overline{)25}$

3 $5\overline{)256}$

4 $3\overline{)742}$

5 $4\overline{)345}$

6 $5\overline{)297}$

7 $7\overline{)2256}$

8 $8\overline{)3468}$

9 $9\overline{)6598}$

Solve these divisions writing your remainder as a decimal. You may need a calculator.

10 $4\overline{)333}$ = 83.25

11 $2\overline{)397}$

12 $4\overline{)505}$

13 $4\overline{)747}$

14 $10\overline{)3897}$

15 $10\overline{)2679}$

16 $5\overline{)366}$

17 $5\overline{)5767}$

18 $5\overline{)5378}$

Number and Algebra Using a calculator

Average mass of Australian coins					
5c	10c	20c	50c	$1	$2
2.83 g	5.65 g	11.3 g	15.55 g	9 g	6.6 g

Calculate the mass of each person's coin collection. Use a calculator.

1 Kiara had two 5c coins, five 10c coins, two 20c coins, one 50c coin, five $1 coins and two $2 coins. What was the total mass of her collection?

2 Darash had one 5c coin, two 10c coins, five 20c coins, four 50c coins, two $1 coins and five $2 coins. What was the total mass of his collection?

3 Celina had five 10c coins, five 50c coins and two $2 coins. What was the total mass of her collection?

4 Abdul had five 5c coins, two 20c coins, one 50c coin and five $1 coins. What was the total mass of his collection?

Number and Algebra

SET 3 Order of operations

Follow the order of operations rules to solve the equations.

> Remember
> [×] and [÷] before
> [+] and [–]
> Do the work in the brackets first.

1. $15 + 18 \div 3 =$
2. $(15 + 18) \div 3 =$
3. $5 + 20 \times 5 =$
4. $(20 + 5) \times 5 =$
5. $21 + 12 \div 3 =$
6. $21 \div 3 + 12 =$
7. $48 \div 12 + 36 =$
8. $48 - 36 \div 12 =$
9. $(36 - 12) + 48 =$
10. $72 \div 8 + 12 =$
11. $(72 + 12) \div 7 =$
12. $72 + 16 \div 8 =$

SET 4 Extension

1. $250 \times 3 + 4$
2. $\frac{7}{10}$ of $50 - 19.5$
3. Average 27, 92, 53, 20
4. Tim travelled 470 km in 5 hours. What was his average speed?
5. Which one is not equivalent: $\frac{3}{5}$, 33% or 0.6?
6. $\frac{5}{8} \times \$72 - \frac{3}{4}$ of 60
7. How much are 12 toys at $10.50 each?
8. Which is larger, $\frac{3}{4}$ or $\frac{3}{5}$?
9. 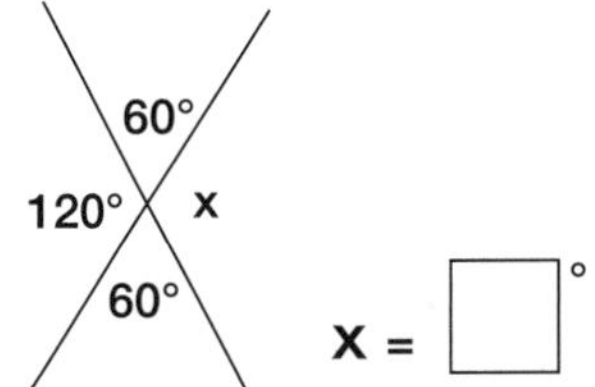

10. A rectangular prism 10 cm × 3 cm × 2 cm has a volume of 60 cm^3. Give the dimensions of 2 other rectangular prisms that have a volume of 60 cm^3.

Mathematical Reasoning

Space Making a map

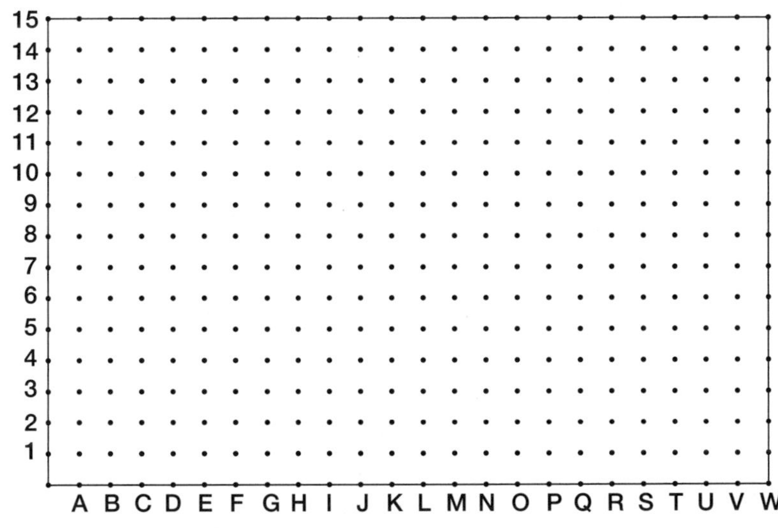

Draw and label the buildings of the school by joining the reference points.

1. School hall: G 1, G 4, O 4, O 1, G 1
2. Senior Classrooms: A 5, A 14, G 14, G 5, A 5
3. Toilets: J 12, J 14, P 14, P 12, J 12
4. Canteen: N 5, N 7, R 7, R 5, N 5
5. Junior Classrooms: Q 1, Q 3, S 3, S 14, U 14, U 1, Q 1

UNIT 35

Number and Algebra

SET 1 Basic

1 49 + 9

2 $200 ÷ 10

3 100 ÷ 5

4 8 ☐ 5 = 1 r 3

5 65 + 35

6 126 ☐ 24 = 102

7 250 – 36

8 $9 \times 7 + 3^2$

9 500 mL = ☐ L

10 Is 33 a prime number?

11 11:57 + 10 minutes

12 Value of 6 in 63 237

13 3 thousands + 36 241

14 Hours in 3 days

15 How many days in a leap year?

16

How many days are there in 4 years?

☐ days

SET 2 Recurring decimals

Solve each division by writing your answer with a decimal remainder. You may need a calculator.

1 $5\overline{)367}$

2 $4\overline{)4865}$

3 $4\overline{)6499}$

4 $5\overline{)777}$

5 $4\overline{)3297}$

6 $8\overline{)2595}$

7 $5\overline{)9353}$

8 $8\overline{)9697}$

9 $8\overline{)2859}$

Solve these divisions writing your remainder with a recurring decimal. You may need a calculator.

10 $3\overline{)394}$

11 $6\overline{)3847}$

12 $9\overline{)3214}$

13 $3\overline{)772}$

14 $6\overline{)3697}$

15 $9\overline{)5105}$

16 $3\overline{)7642}$

17 $6\overline{)1529}$

18 $9\overline{)4391}$

Statistics and Probability Data exploration

Recently 50 tickets were sold to the Youth Club Freak Night. Use the illustration showing the age of each ticket purchaser to construct a tally and frequency table to represent the ages of the people attending. It has been started for you.

(14) (14) (14) (15) (15) (15) (15)
(15) (16) (16) (16) (16) (16) (16)
(16) (17) (17) (17) (17) (17) (17)
(17) (17) (17) (17) (17) (17) (17)
(18) (18) (18) (18) (18) (18) (18)
(18) (18) (18) (18) (18) (18) (18)
(18) (19) (19) (19) (19) (19) (19) (19)

Age	Tally	Frequency
14	|||	3
15		
16		
17		
18		
19		

Number and Algebra

SET 3 Order of operations

Substitute the answers in the boxes into the number sentences to see which number is the missing number. Circle your choice.

		Answers		
1	(5 + 4) × ☐ = 54	5	6	7
2	6 + 3 × ☐ = 21	3	4	5
3	☐ ÷ 6 + 8 = 11	18	19	20
4	(7 + ☐) × 6 = 90	7	8	9
5	42 ÷ 7 ÷ ☐ = 3	6	3	2
6	☐ × 4 ÷ 5 = 20	20	25	30
7	(19 – ☐) ÷ 7 = 2	9	7	5
8	36 – 6 × ☐ = 12	5	9	4

Mathematical Reasoning

9 Below the problem are two number sentences. Tick the one you think solves the problem.

Problem: David earns $500 per week and pays $125 tax out of that money. How much 'take home pay' does he earn per year?

($500 – $125) × 52 = $19 500 ☐

$500 × 52 – $125 = $25 875 ☐

SET 4 Extension

1 25 × 20 = 500 True or false?

2 Perimeter of a regular nonagon with sides of 24 cm

3 Average of 3.75, 4.25, 5.9 and 2.1

4 How many kilograms in 14.75 tonnes?

5 3.5 × 10

6 Round off and estimate 119.8 × 6.9.

7 How many 125 g packets of peanuts in 4 kg?

8 Round then divide 2499 by 5.

9

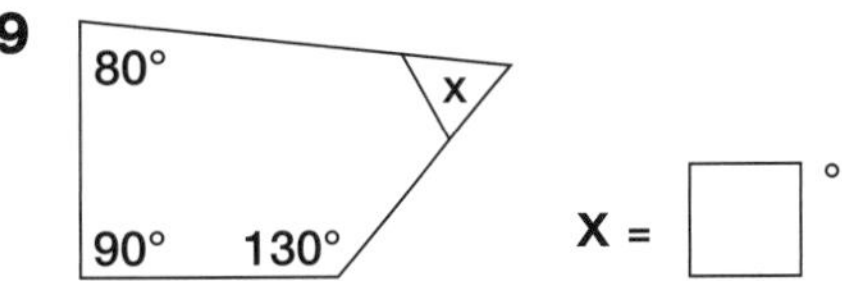

10 Parallel lines never meet. True or false?

11 Write 11:27 pm in 24-hour time.

12 3.5 hours at an average speed of 80 km per hour = ☐ km

13 Minutes in 1.5 days

14 Which does not fit: 0.6, 60%, $\frac{6}{100}$ or $\frac{6}{10}$?

15 Apply 6 × (■ + 3) – 7 = ●

■	3	1	5	20	40	72
●						

Measurement Mass

1 Complete the table to show how the same measurement can be expressed in 3 ways.

Kilograms	5467 kg				7859 kg		
Tonnes and kilograms		6 t 381 kg		4 t 607 kg		3 t 555 kg	
Tonnes			9.567 t				8.016 t

Use the greater than (>) or less than (<) signs to complete these comparisons.

2 3.789 t ☐ 3897 kg **3** 5679 kg ☐ 5.6 t **4** 5 t 452 kg ☐ 5.5 t **5** 5900 kg ☐ 5.899 t

Maths helpers

Length

10 millimetres (mm) = 1 centimetre (cm)

100 centimetres (cm) = 1 metre (m)

1000 metres (m) = 1 kilometre (km)

Mass

1000 grams (g) = 1 kilogram (kg)

1000 kg = 1 tonne (t)

Capacity

1000 millilitres (mL) = 1 litre (L)

Time

60 seconds = 1 minute

60 minutes = 1 hour

24 hours = 1 day

7 days = 1 week

14 days = 1 fortnight

12 months = 1 year

52 weeks = 1 year

365 days = 1 year

366 days = 1 leap year

10 years = 1 decade

100 years = 1 century

Months of the year

Thirty days has September, April, June and November. All the rest have thirty-one, except February alone, which has twenty-eight days clear and twenty-nine days each leap year.

Seasons

Summer: December, January, February

Autumn: March, April, May

Winter: June, July, August

Spring: September, October, November

Roman numerals

1 = I

2 = II

3 = III

4 = IV

5 = V

6 = VI

7 = VII

8 = VIII

9 = IX

10 = X

20 = XX

30 = XXX

40 = XL

50 = L

60 = LX

70 = LXX

80 = LXXX

90 = XC

100 = C

500 = D

1000 = M

Multiplication facts

×	0	1	2	3	4	5	6	7	8	9	10
0	0	0	0	0	0	0	0	0	0	0	0
1	0	1	2	3	4	5	6	7	8	9	10
2	0	2	4	6	8	10	12	14	16	18	20
3	0	3	6	9	12	15	18	21	24	27	30
4	0	4	8	12	16	20	24	28	32	36	40
5	0	5	10	15	20	25	30	35	40	45	50
6	0	6	12	18	24	30	36	42	48	54	60
7	0	7	14	21	28	35	42	49	56	63	70
8	0	8	16	24	32	40	48	56	64	72	80
9	0	9	18	27	36	45	54	63	72	81	90
10	0	10	20	30	40	50	60	70	80	90	100

Addition facts

+	2	3	4	5	6	7	8	9	10	11	12
2	4	5	6	7	8	9	10	11	12	13	14
3	5	6	7	8	9	10	11	12	13	14	15
4	6	7	8	9	10	11	12	13	14	15	16
5	7	8	9	10	11	12	13	14	15	16	17
6	8	9	10	11	12	13	14	15	16	17	18
7	9	10	11	12	13	14	15	16	17	18	19
8	10	11	12	13	14	15	16	17	18	19	20
9	11	12	13	14	15	16	17	18	19	20	21
10	12	13	14	15	16	17	18	19	20	21	22
11	13	14	15	16	17	18	19	20	21	22	23
12	14	15	16	17	18	19	20	21	22	23	24

UNIT 1 Number and Algebra

SET 1

1 32
2 48
3 17
4 ×
5 ÷
6 81
7 9
8 ×
9 55
10 24
11 9
12 869c
13 72
14 1, 3, 5, 9, 15, 45
15 $6.50

SET 2

1 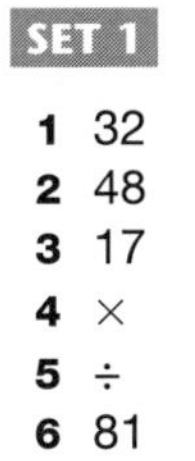

2

3 64 928
4 52 818 52 819
5 $193
6 47 000
7 157 695
8 252 340
9 349 hundreds
10 443 186
11 214 900 214 899
12 25 133
13 975 970

SET 3

1 86
2 143
3 148
4 135
5 177
6 464
7 867
8 1156
9 47
10 32
11 34
12 112
13 201
14 614

	Question	Rounded to 100	Approximate answer
15	395 + 206	400 + 200	600
16	591 − 298	600 − 300	300
17	513 + 387	500 + 400	900
18	785 − 589	800 − 600	200
19	372 + 329	400 + 300	700
20	882 − 286	900 − 300	600

SET 4

1 9.4
2 800
3 8000
4 20
5 8
6 6.1
7 $8.25
8 110
9 90
10 2000
11 True
12 13°C
13 4 km
14 8000
15 $1.35
16 15

Space

1 obtuse angle
2 reflex angle
3 acute angle
4 right angle

Statistics and Probability

1 12
2 28
3 22
4 14
5 12

UNIT 2 Number and Algebra

SET 1

1 11
2 0
3 7
4 250
5 49
6 12 February
7 7
8 $5
9 71
10 Winter
11 ÷
12 4
13 20
14 7000
15 75 cm

SET 2

1 9 cm^2
2 16 cm^2
3 25 cm^2
4 4 cm^2
5 36
6 49
7 64
8 100
9 1
10 81

SET 3

1 −10°C
2 −15°C
3 −5°C
4 8
5 −2
6 2
7 3
8 −4
9 −4
10 0
11 3

SET 4

1 90
2 23%
3 6 ones
4 35
5 $72
6 6
7 63 200
8 $31.00
9 True
10 803
11 9 × 6 = 54
12 3 × 2 = 6
13 9 ÷ 3 = 3
14 9 × 3 + 6 = 33
15 72 ÷ 9 × 6 = 48
16 54 ÷ 6 = 3 × 3

Space

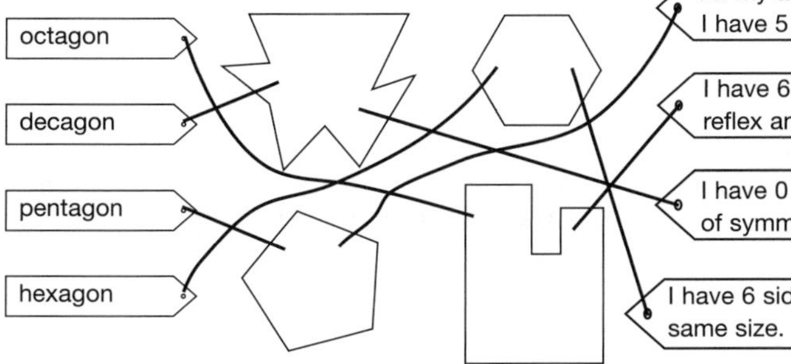

Measurement

1 170 mm
2 200 mm
3 120 mm
4 100 mm
5 100 mm
6 120 mm

Answers

UNIT 3 Number and Algebra

SET 1

1 15
2 12
3 9
4 ×
5 –
6 9
7 6
8 ÷
9 28
10 21
11 7
12 70
13 Yes
14 1216
15 250 m

SET 2

1 True
2 True
3 False
4 True
5 False
6 True
7 True
8 True
9 True
10 False
11 True
12 True
13 True
14 True
15 5 is a factor if 0 or 5 is in the units column

SET 3

1 $\frac{10}{100}$, 0.10, 10%
2 $\frac{25}{100}$, 0.25, 25%
3 $\frac{7}{10}$, 0.7, 70%
4 $\frac{20}{100}$, 0.2, 20%
5 $\frac{1}{2}$, 0.5, 50%
6 $\frac{1}{4}$, 0.25, 25%
7 $\frac{3}{4}$, 0.75, 75%
8 $\frac{27}{100}$, 0.29, 30%
9 0.33, 35%, $\frac{53}{100}$
10 9%, 0.9, $\frac{99}{100}$
11 49%, $\frac{1}{2}$, 0.54
12 4%, 0.21, $\frac{3}{10}$
13 $\frac{9}{100}$, 90%, 0.95
14 0.03, 7%, $\frac{70}{100}$

SET 4

1 70 000
2 1%
3 180 cm
4 4
5 43
6 3158
7 76
8 35°C
9 7300
10 29 000
11 60
12 7.5 cm
13 4
14 $\frac{24}{40}$
15 0.15, 15% or 15 hundredths
16 $31.50

Space

	Name	Faces	Vertices	Edges
1	Cylinder	3	0	2
2	Rectangular pyramid	5	5	8
3	Rectangular prism	6	8	12
4	Triangular prism	5	6	9

Measurement

	L	W	A
1	4 cm	3 cm	12 cm²
2	3 cm	2 cm	6 cm²
3	7 cm	2 cm	14 cm²

UNIT 4 Number and Algebra

SET 1

1 45
2 7
3 22
4 16
5 –
6 ÷
7 5
8 +
9 $7
10 5
11 1, 3, 9
12 Yes
13 No
14 700
15 $11

SET 2

1 323
2 142
3 35
4 26
5 109
6 93
7 $8
8 18
9 45
10 51
11 106
12 $132
13 54
14 31

SET 3

1 Rule: Multiply by 6

2	4	6	8	10	12	14	16	18
12	24	36	48	60	72	84	96	108

2 Rule: Subtract 12

120	108	96	84	72	60	48	36	24
108	96	84	72	60	48	36	24	12

3 Rule: Divide by 8

96	88	80	72	64	56	48	40	32
12	11	10	9	8	7	6	5	4

4 Rule: Add 5^2

10	20	30	40	50	60	70	80	90
35	45	55	65	75	85	95	105	115

5 Hands on.

SET 4

1 75
2 9
3 700
4 5704, 5804
5 6
6 243
7 $19
8 140
9 3
10 $6.80
11 250
12 $\frac{1}{4}, \frac{1}{3}, \frac{1}{2}, \frac{3}{4}$
13 10:23
14 $6.48
15 $14.45
16 $6.50
17 25 000 (50 × 500)

Measurement

	Shape	L	W	H	Volume
1	A	4	2	3	24 cm³
2	B	5	3	4	60 cm³
3	C	6	3	4	72 cm³
4	D	7	4	5	140 cm³
5	E	11	2	2	44 cm³

Statistics and Probability

1st toss	2nd toss	3rd toss	Outcome
Heads	Heads	Heads	H H H
		Tails	H H T
	Tails	Heads	H T H
		Tails	H T T
Tails	Heads	Heads	T H H
		Tails	T H T
	Tails	Heads	T T H
		Tails	T T T

UNIT 5 Number and Algebra

SET 1

1 24
2 36
3 9
4 ×
5 ÷
6 64
7 8
8 ×
9 60
10 23
11 6
12 961c
13 90
14 1, 3, 9, 27
15 216, 432

SET 2

1 27
2 270
3 2700
4 2700
5 630
6 5600
7 1500
8 4200
9 90
10 200
11 360
12 20
13 350
14 800
15 1500
16 50
17 140
18 210
19 330
20 176
21 216

SET 3

1 0.3
2 0.5
3 0.15
4 0.24
5 0.52
6 0.44
7 7.42
8 6.12
9 22.95
10 63.56
11 307.2
12 258.19

SET 4

1 348
2 $1\frac{3}{4}$
3 6000
4 6894 6884
5 100°C
6 $196
7 40%, 0.47, $\frac{1}{2}$, $\frac{3}{5}$
8 360
9 $46.72
10 306 km
11 95°
12 10 000 m²
13 90
14 10
15 $48.75
16 920 mm

Space

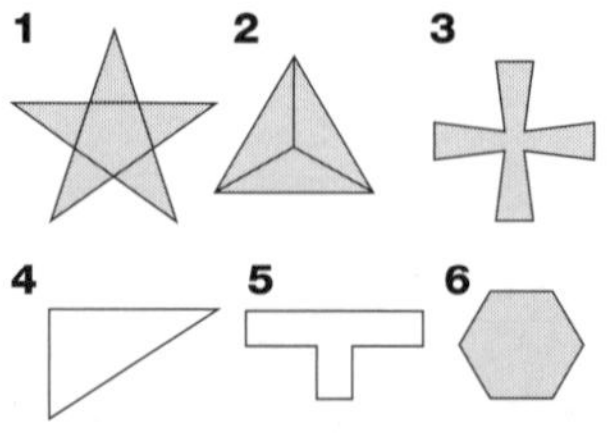

Statistics and Probability

1

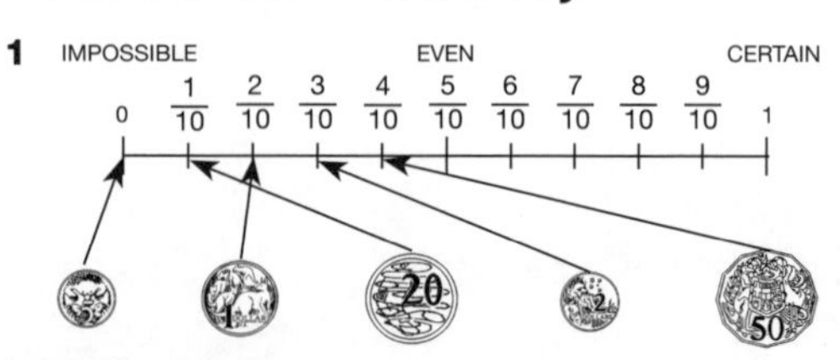

2 Very unlikely

UNIT 6 Number and Algebra

SET 1

1 –
2 24
3 7
4 6
5 160
6 +
7 ÷
8 63
9 18
10 6000
11 1, 2, 4, 5, 10, 20
12 1423
13 No
14 210
15 10

SET 2

1 35°C
2 –5°C
3 5°C
4 15°C
5 5°C
6 10°C
7 –15°C
8 –15°C
9 –10°C
10 –30°C

SET 3

1 True
2 True
3 True
4 True
5 False
6 True
7 False
8 False
9 True
10 True
11 $1\frac{1}{3}$
12 $1\frac{2}{5}$
13 $2\frac{1}{4}$
14 $1\frac{2}{6}$
15 $2\frac{1}{5}$
16 $2\frac{1}{2}$

Possible solutions
17 $\frac{5}{2}$, $\frac{10}{4}$
18 $\frac{5}{4}$, $\frac{10}{8}$
19 $\frac{4}{3}$, $\frac{8}{6}$
20 $\frac{13}{6}$, $\frac{26}{12}$

SET 4

1 105
2 $7.08
3 400 000
4 37
5 0.2, $\frac{1}{4}$, 0.31, 35%
6 $28
7 $19
8 500 m
9 20
10 600
11 $42.82
12 700 m
13 $\frac{36}{100}$
14 $\frac{15}{8}$ or $1\frac{7}{8}$
15 9.3
16 65
17 7.481 5.373 2.299
18 7.434 5.326 2.252
19 7.931 5.823 2.749
20 7.432 5.324 2.250

Statistics and Probability

1 Fish
2 25
3 Bird
4 Cat and Bird

Measurement

1 8:55 am
2 Stuart Street
3 30 min
4 24 min
5 10:00 am

UNIT 7 Number and Algebra

SET 1

1 8
2 35
3 16
4 49
5 18
6 460
7 42
8 $3
9 ÷
10 9
11 autumn
12 No
13 9
14 32
15 192

SET 2

1 $\frac{5}{8}$
2 $\frac{8}{10}$
3 $\frac{9}{10}$
4 $\frac{7}{8}$
5 $\frac{12}{8}$ or $1\frac{4}{8}$
6 $\frac{6}{4} = 1\frac{2}{4}$
7 $\frac{14}{10} = 1\frac{4}{10}$
8 $\frac{7}{5} = 1\frac{2}{5}$
9 $\frac{15}{10}$ or $1\frac{5}{10}$
10 $\frac{13}{12}$ or $1\frac{1}{12}$
11 $\frac{6}{10}$
12 $\frac{5}{8}$
13 $\frac{4}{10}$
14 $\frac{1}{5}$
15 $\frac{1}{4}$
16 $\frac{3}{10}$
17 $\frac{1}{8}$
18 $\frac{5}{10}$
19 $\frac{1}{5}$
20 $\frac{6}{10}$

SET 3

1 ones
2 tens
3 hundreds
4 tenths
5 hundredths
6 thousand
7 hundredths
8 thousandths
9 7521 g
10 5425 mL
11 1785 mm
12 6301 kg

SET 4

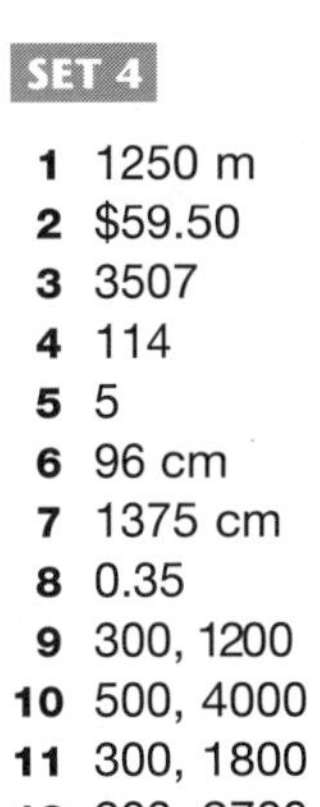

1 1250 m
2 $59.50
3 3507
4 114
5 5
6 96 cm
7 1375 cm
8 0.35
9 300, 1200
10 500, 4000
11 300, 1800
12 900, 2700
13 500, 2500
14 900, 8100

Space

Number and Algebra

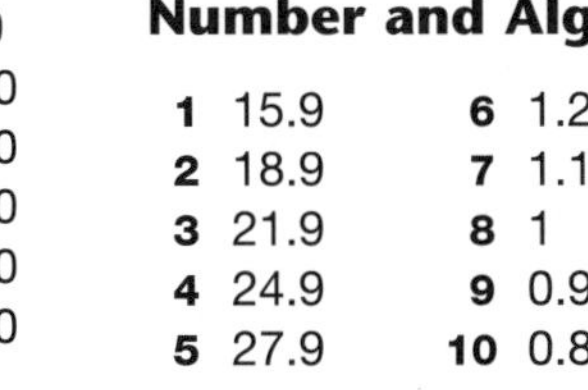

1 15.9
2 18.9
3 21.9
4 24.9
5 27.9
6 1.2
7 1.1
8 1
9 0.9
10 0.8

UNIT 8 Number and Algebra

SET 1

1 8
2 35
3 24
4 27
5 +
6 ÷
7 30
8 3
9 ×
10 400
11 $\frac{7}{10}$
12 1, 3, 5, 15
13 Yes
14 15
15 1 L

SET 2

1 150 643
2 174 727
3 51 109
4 29 926
5 $46 435
6 $56 107
7 $91 564
8 $38 375

SET 3

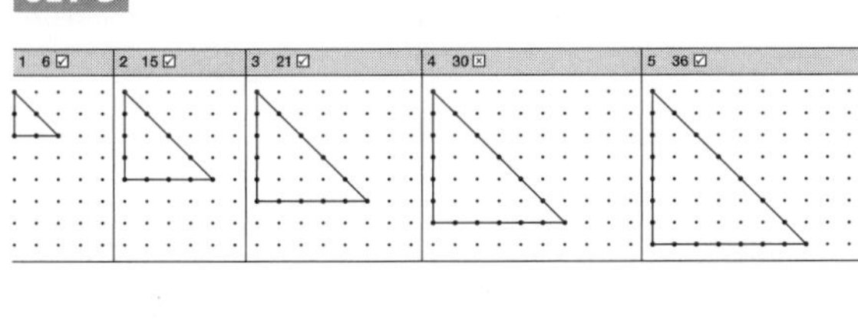

SET 4

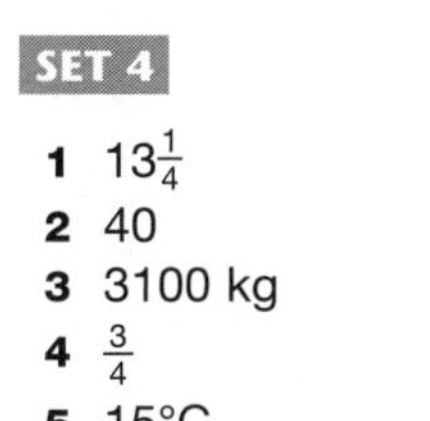

1 $13\frac{1}{4}$
2 40
3 3100 kg
4 $\frac{3}{4}$
5 15°C
6 1 226 000
7 $2112
8 4
9 60°
10 3.5, 3.75, $3\frac{9}{10}$, $4\frac{1}{4}$
11 $32
12 6
13 33

Statistics and Probability

1 25 kg
2 45 kg
3 6 and 7 years
4 15 kg
5 20 kg
6 $37\frac{1}{2}$ kg

Space

Possible solutions.

1 60°
2 40°
3 120°
4 90°

Answers

UNIT 9 Number and Algebra

SET 1

1. 49
2. ×
3. ÷
4. 64
5. 28
6. 45
7. 7
8. 63
9. 38
10. 3752 c
11. No
12. 165
13. 30%
14. $3.30
15. Saturday

SET 2

1. $1050
2. 367 km
3. 693
4. 327
5. $588
6. $625

SET 3

1. composite
2. prime
3. composite
4. composite
5. composite
6. composite
7. composite
8. composite
9. composite
10. composite
11. prime
12. composite

13	18	1, 2, 3, 6, 9, 18
14	21	1, 3, 7, 21
15	32	1, 2, 4, 8, 16, 32
16	56	1, 2, 4, 7, 8, 14, 28, 56

SET 4

1. 370 cm
2. 31 004
3. 7
4. 125
5. 20 000
6. 105
7. 33
8. 0.51
9. 2350 m
10. 290 006
11. $2
12. $34

Statistics and Probability

Chance

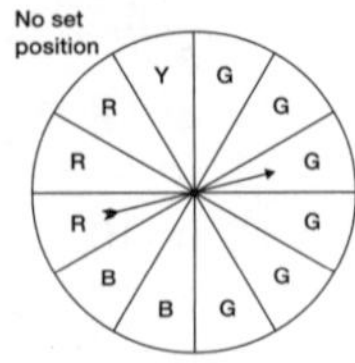

Statistics and Probability

1. 8°C
2. 8°C
3. 8°C
4. 0°C
5. 0°C

UNIT 10 Number and Algebra

SET 1

1. 16
2. 56
3. 15
4. +
5. ×
6. 25
7. 7
8. –
9. 42
10. 9
11. 1324 c
12. 27
13. No
14. 3175 mL
15. 4
16. 12

SET 2

1. 58 990
2. 105 303
3. 50 877
4. 107 161
5. 82 413
6. 335

SET 3

1. $\frac{2}{4}$
2. $\frac{5}{10}$
3. $\frac{6}{10}$
4. $\frac{2}{8}$
5. $\frac{3}{8}$
6. $\frac{13}{10} = 1\frac{3}{10}$
7. $\frac{7}{5} = 1\frac{2}{5}$
8. $\frac{11}{10} = 1\frac{1}{10}$
9. $\frac{17}{10} = 1\frac{7}{10}$
10. $\frac{5}{4} = 1\frac{1}{4}$
11. $\frac{7}{4} = 1\frac{3}{4}$
12. $\frac{18}{10} = 1\frac{8}{10}$
13. $\frac{11}{5} = 2\frac{1}{5}$
14. $\frac{21}{10} = 2\frac{1}{10}$
15. $\frac{17}{8} = 2\frac{1}{8}$

SET 4

1. 90
2. True
3. 475 cm
4. 332
5. 10
6. 11.16
7. 8
8. 150
9. 100°C
10. 105 226
11. 37 190
12. 1
13. 80%
14. 41 404 (207 020 ÷ 5)

Space

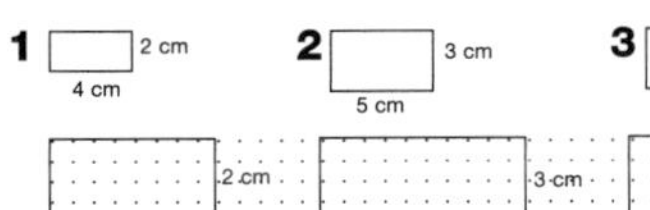

Measurement

1. 647 km
2. 755 km
3. 766 km
4. 741 km
5. 1402 km

UNIT 11 Number and Algebra

SET 1

1. 8 r 1
2. 15
3. W
4. 39
5. 730
6. 61
7. 20 cm
8. 549
9. 25
10. 0.21
11. 0
12. 3 r 2
13. 61
14. 678
15. 9
16. 15

SET 2

1. 0
2. –4
3. –4
4. –10
5. –$30
6. 10kg
7. –5°C
8. –$30
9. –50

SET 3

1. 18
2. 6
3. 9
4. 3
5. 4
6. 12
7. 30
8. 20
9. 30
10. 40
11. 60
12. 60
13. 30
14. 80
15. 50
16. 160
17. 80
18. 600
19. 50
20. 20

SET 4

1. $\frac{22}{40}$
2. 1710
3. 60 cm^3
4. 5250 kg
5. 0°C
6. 320 m^2
7. $19
8. 5
9. 40°
10. $34.50
11. 4350 m
12. 637 430
13. 8
14. 74

Statistics and Probability

1. 1 pm
2. 4
3. 24
4. 10

Measurement

1. 10
2. 20
3. 4
4. 100
5. 50

UNIT 12 Number and Algebra

SET 1

1 180
2 120
3 –
4 –
5 ×
6 7 r 1
7 +
8 0
9 147
10 $\frac{9}{10}$
11 60 000
12 No
13 True
14 3270 mL
15 31
16 $13.50

SET 2

1 41
2 161
3 38
4 26
5 105
6 64
7 58
8 188
9 9.6
10 3.2
11 10
12 6
13 11
14 $357

SET 3

1 10%
2 0.2
3 0.5
4 23%
5 0.99
6 0.2
7 0.75
8 100%
9 $25
10 $10
11 $10
12 $4
13 $10
14 $25

SET 4

1 0.37
2 $\frac{1}{5}$, 30%, 0.31, $\frac{9}{10}$
3 $2\frac{3}{10}$
4 $65
5 7
6 $\frac{5}{8}$
7 5.5
8 10
9 14 cm
10 $3.90
11 $3
12 $21
13 4
14 Sally: 1.63 m
Anna: 1.53 m

Space

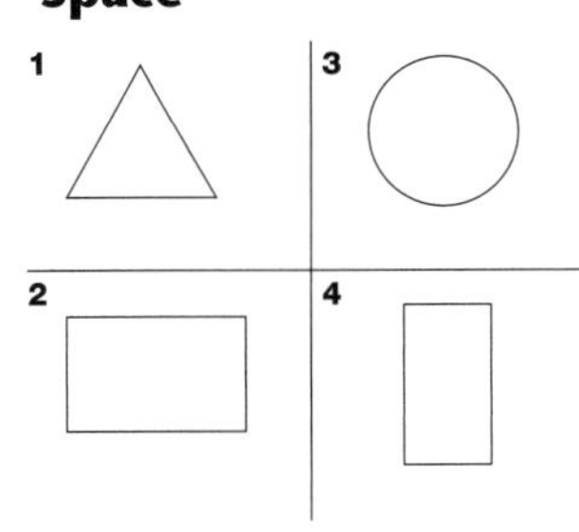

Statistics and Probability

1 Y R Y R R Y Y R
R R Y Y Y R R Y
R Y R Y Y Y R R

2 7 is the more likely score. There are 6 combinations of scores that have a sum of 7 but only 3 that have a sum of 10.

UNIT 13 Number and Algebra

SET 1

1 +
2 –
3 36
4 9
5 49
6 16
7 39
8 43
9 No
10 Yes
11 2 r 1
12 30 000
13 $1.35
14 30
15 4
16 695

SET 2

1 4.71
2 5.45
3 1.12
4 4.13
5 2.46
6 1.23
7 53.01
8 24.13
9 22.51
10 5.401
11 22.03
12 2.273

SET 3

1 30 000 + 7000 + 400 + 20 + 3
2 80 000 + 5000 + 600 + 10 + 6
3 20 000 + 5000 + 200 + 9
4 100 000 + 6000 + 10 + 5
5 6 000 000 + 40 000 + 1000 + 500
6 68 499
7 154 347
8 265 358
9 71 358
10 54 812 405
11 480 207
12 4 310 100
13 5 541 040

SET 4

1 19.5 m
2 $3.85
3 $145
4 $1\frac{4}{10}$, 1.49, 150%, 1.53
5 125
6 2 soccer fields
7 175 cm
8 24.5 cm^2
9 750 cm
10 10 000 cm^2
11 35°
12 $4.10
13 2
14 5750 mL
15 Cube

Statistics and Probability

1 Hands on.
Scale on graph is misleading. The price has only increased by $25.

Measurement

1 50 km
2 10 km
3 30 km
4 25 km
5 35 km
6 45 km

UNIT 14 Number and Algebra

SET 1

1 –
2 21
3 15
4 +
5 6
6 75
7 15
8 13
9 $2.65
10 40
11 Yes
12 8
13 14
14 51 km
15 $48

SET 2

1 5368
2 6735
3 20 865
4 14 046
5 8432
6 36 281
7 14 872
8 30 245
9 20 196
10 $42 632
11 $82 908

SET 3

1 $\frac{2}{4}$
2 $\frac{2}{8}$
3 $\frac{3}{12}$
4 $\frac{2}{6}$
5 $\frac{4}{8}$
6 $\frac{5}{10}$
7 $\frac{4}{12}$
8 $\frac{2}{10}$
9 $\frac{3}{15}$
10 $\frac{6}{10}$
11 $\frac{4}{6}$
12 $\frac{6}{8}$

SET 4

1 $180
2 7
3 $16.20
4 $3.00
5 About 12 000
6 1315
7 740
8 $\frac{4}{9}$
9 $\frac{1}{12}$
10 5
11 15
12 8
13 $9.90
14 Park = 500 m
Preschool = 400 m

Number and Algebra

Number	Prime or composite	Explain why
17	Prime	It can only be divided by itself (17) or by 1
51	Composite	It can be divided by 3 and 17
39	Composite	It can be divided by 3 and 13
85	Composite	It can be divided by 5 and 17
43	Prime	It can only be divided by 43 or by 1

Space

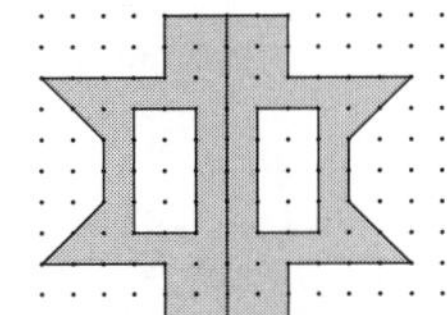

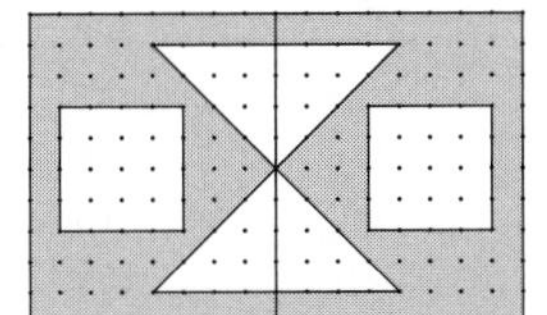

Answers

UNIT 15 Number and Algebra

SET 1

1 22nd
2 $11\frac{1}{2}$
3 3 r 1
4 ÷
5 90
6 4th
7 2 May
8 7
9 35c
10 48c
11 20
12 25
13 $6.50
14 64
15 $12.50

SET 2

1 0 1 2 3 4
$\frac{1}{2}$ $1\frac{1}{4}$ $1\frac{3}{4}$ $2\frac{1}{3}$ $3\frac{7}{8}$
2 $\frac{1}{4}$ $1\frac{1}{4}$ $2\frac{2}{4}$ $3\frac{1}{8}$ $3\frac{7}{8}$
3 $3\frac{1}{2}$, $4\frac{1}{2}$, $5\frac{1}{2}$, $6\frac{1}{2}$, $7\frac{1}{2}$, $8\frac{1}{2}$
4 $7\frac{1}{4}$, $7\frac{1}{2}$, $7\frac{3}{4}$, 8, $8\frac{1}{4}$, $8\frac{1}{2}$
5 $8\frac{1}{3}$, 8, $7\frac{2}{3}$, $7\frac{1}{3}$, 7, $6\frac{2}{3}$
6 $6\frac{1}{4}$, $6\frac{3}{4}$, $7\frac{1}{4}$, $7\frac{3}{4}$, $8\frac{1}{4}$, $8\frac{3}{4}$
7 7, $7\frac{3}{4}$, $8\frac{1}{2}$, $9\frac{1}{4}$, 10, $10\frac{3}{4}$
8 $5\frac{3}{5}$, $6\frac{1}{5}$, $6\frac{4}{5}$, $7\frac{2}{5}$, 8, $8\frac{3}{5}$
9 $7\frac{7}{10}$, $8\frac{3}{10}$, $8\frac{9}{10}$, $9\frac{5}{10}$, $10\frac{1}{10}$, $10\frac{7}{10}$

SET 3

1 2000 + 4000 = 6000, 6054
2 9000 – 3000 = 6000, 5916
3 6000 + 4000 = 10 000, 10 039
4 4000 + 2000 = 6000, 6004
5 8000 + 1000 = 9000, 8986
6 6000 – 4000 = 2000, 1932
7 20 000
8 42 000
9 81 000
10 30 000
11 16 000
12 32 000
13 24 000

SET 4

1 1915
2 $5.50
3 $8.10
4 116
5 105
6 90 km/h
7 5250 m
8 $1.90
9 0.5, 5.0, 15.0, 5^2
10 $146
11 135° (225° if measured anti-clockwise.)
12 The volume increases 8 times.

Space

1 18 km
2 9 km
3 12 km
4 $63
5 $31.50
6 $42

Statistics and Probability

1 0.2
2 0.2
3 0.1
4 0.4
5 0.1

UNIT 16 Number and Algebra

SET 1

1 81
2 61
3 38
4 $2.48
5 165
6 2025
7 $250
8 3.97
9 $5.40
10 0.19
11 30
12 1483
13 600 sec
14 12
15 9
16 90

SET 2

1 2043
2 369
3 1123
4 658
5 387
6 651
7 $1376\frac{3}{5}$
8 $1825\frac{2}{4}$ or $1825\frac{1}{2}$
9 $1952\frac{1}{3}$
10 $1232\frac{2}{6}$ or $1232\frac{1}{3}$
11 $825\frac{3}{8}$
12 $1452\frac{4}{5}$
13 364
14 145

SET 3

1 $\frac{3}{6}$
2 $\frac{3}{12}$
3 $\frac{5}{12}$
4 $\frac{4}{6}$
5 $\frac{7}{12}$
6 $\frac{9}{6} = 1\frac{3}{6}$
7 $\frac{7}{6} = 1\frac{1}{6}$
8 $\frac{15}{12} = 1\frac{3}{12}$
9 $\frac{19}{12} = 1\frac{7}{12}$
10 $\frac{13}{12} = 1\frac{1}{12}$

SET 4

1 $3 × 20 = $60
2 $240
3 $1\frac{4}{8} = 1\frac{1}{2}$
4 $9
5 4.88
6 $30
7 75 cm
8 33
9 54
10 37
11 1, 1.5, $1\frac{3}{5}$, 1.61
12 $10.10
13 30
14 700 kg
15 $\frac{1}{15}$
16 224 cm^3
17 $1\frac{1}{2}$, $1\frac{1}{8}$, $1\frac{2}{5}$, $1\frac{1}{2}$, $1\frac{2}{3}$, $1\frac{1}{2}$, $1\frac{1}{4}$, $2\frac{1}{3}$

Number and Algebra

1

2

Hexagons	1	2	3	4	5	6	7
Sides	6	12	18	24	30	36	42

3 10 × 6 = 60 sides

Measurement

1 120 m^3
2 60 m^3

UNIT 17 Number and Algebra

SET 1

1 42
2 63
3 29
4 4
5 ×
6 ÷
7 80
8 8
9 99
10 32
11 52
12 Yes
13 No
14 1887
15 4
16 10

SET 2

1 400
2 465
3 1260
4 3648
5 2080
6 2475
7 10 314
8 18 072
9 24 948
10 20 382
11 16 994
12 40 176

SET 3

1 $8, $16, $24, $32, $40, $48, $56
2 $12, $24, $36, $48, $60, $72, $84
3 $20, $40, $60, $80, $100, $120, $140
4 $4.50, $9, $13.50, $18, $22.50, $27, $31.50
5 Hands on.

SET 4

1 1525
2 45 months
3 $\frac{1}{8}$
4 102
5 6
6 No
7 30°
8 True
9 75 L
10 2350
11 77.7
12 1 026 231
13 7100 mL
14 None
15 Rectangular prism
16 40

Number and Algebra

1 $967.89
2 $1519.72
3 $578.95
4 $224.86
5 12780.90 kg
6 114.722 km
7 1570.83 m
8 68.235 kg

Measurement

1 5 2 25 3 8

UNIT 18 Number and Algebra

SET 1

1 48
2 63
3 45
4 ÷
5 6
6 +
7 Yes
8 22
9 30
10 1, 2, 3, 6, 9, 18
11 3 745
12 4
13 210
14 701
15 80
16 $84

SET 2

1 5
2 6
3 6
4 3
5 6
6 14
7 16
8 12
9 25
10 30
11 False
12 False
13 True
14 True
15 False
16 Book B ($2)

SET 3

1 5908
2 8347
3 4542
4 11 809
5 5332
6 11 983
7 $3999
8 $14 729

SET 4

1 360 min
2 $50
3 $9
4 60°
5 54
6 900 000
7 8
8 90 km/h
9 32
10 6
11 6 000 000
12 2257
13 $140
14 1
15 1.4 m

Space

1 To the right

2 Directly below

3 Directly above

Measurement

Cubic centimetres	Millilitres
3 cm^3	3 mL
6 cm^3	6 mL
12 cm^3	12 mL
24 cm^3	24 mL
36 cm^3	36 mL
24 cm^3	24 mL

Cubic centimetres	Millilitres	Litres
200 cm^3	200 mL	
250 cm^3	250 mL	
400 cm^3	400 mL	
1000 cm^3	1000 mL	1 L
2000 cm^3	2000 mL	2 L
3000 cm^3	3000 mL	3 L

UNIT 19 Number and Algebra

SET 1

1 56
2 59
3 4000
4 66
5 ÷
6 12
7 60
8 9
9 57 631
10 $4.80
11 700
12 Yes
13 No
14 5
15 65c

SET 2

1 5.92
2 5.22
3 24.54
4 $28.90
5 $2.55
6 $3
7 8.989 km
8 $54.40
9 $12.14
10 $19.20
11 $4.55
12 1.256 m
13 $1.55
14 $0.95
15 $1.75
16 $1.95

SET 3

1 $7.44
2 $45.69
3 $76.15
4 $172.75
5 $241.85
6 $142.00

SET 4

1 11
2 103°
3 90
4 25% of 1000
5 9 hundredths
6 159 000
7 30
8 16, 36
9 300 × 20 = 6000
10 $37.10
11 248 008
12 $150
13 $\frac{6}{25}$
14 510 mm
15 $1\frac{5}{8}$
16 288

Number and Algebra

1 (+2)

0	1	2	3	4	5
2	3	4	5	6	7

2 (−1)

6	5	4	3	2	1
5	4	3	2	1	0

Measurement

1 2118 / 9:18
2 0405 / 4:05
3 1713 / 5:13
4 1344 / 1:44
5 1142 / 11:42

UNIT 20 Number and Algebra

SET 1

1 56
2 180°
3 5
4 +
5 11
6 1642
7 50
8 1, 3, 7, 21
9 $6.25
10 12
11 2032
12 6 hundredths
13 ÷
14 16 025 g
15 71

SET 2

1 3100
2 1800
3 860
4 4500
5 1500
6 210
7 1400
8 $99
9 $200
10 $60
11 3000 g

SET 3

Possible solutions:
1 ($150 × 5) + ($175 × 6) = $1800
2 (5 × $3.45) + (7 × $2.75) + (3 × $4.50) = $50
3 (640 ÷ 4) × 13 = $2080
4 (500 ÷ 100) × 25 = $125
5 $56 800 × 0.25 = $14 200

SET 4

1 85
2 $1\frac{27}{100}$
3 $2.20
4 $31.20
5 202 min
6 288 L
7 411 000
8 36
9 8 L
10 18
11 $34
12 Hands on

Statistics and Probability

1 Summer
2 Summer
3 Spring
4 Approximately $2250

Space

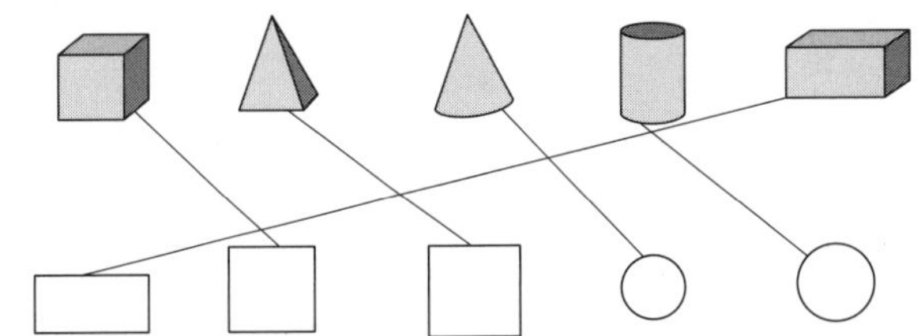

Answers

UNIT 21 Number and Algebra

SET 1

1 32
2 10:45
3 250
4 9
5 63
6 951
7 90°C
8 14.7
9 5.9
10 172
11 30
12 81
13 67 km
14 37
15 $3.50

SET 2

1

1	Balance	
2		$1998
3	= C2 – B3	$1748
4	= C3 – B4	$1483
5	= C4 – B5	$1243
6	= C5 – B6	$ 863
7	= C6 – B7	$ 628
8	= C7 – B8	$ 498
9	= C8 – B9	$ 268
10	= C9 – B10	$ 8

2 645
3 715

SET 3

1 $\frac{2}{10}$
2 $\frac{3}{30}$
3 $\frac{4}{24}$
4 $\frac{6}{15}$
5 $\frac{15}{20}$
6 $\frac{8}{12}$
7 $\frac{2}{5}$
8 $\frac{3}{5}$
9 $\frac{2}{3}$
10 $\frac{2}{3}$
11 $\frac{3}{5}$

SET 4

1 19 × 4 = 76
2 $84.20
3 390
4 $13.50
5 12 000
6 121.67
7 $1.30
8 $1\frac{3}{10}$, 137%, 1.4, 1.45
9 54
10 827 000
11 1.27
12 10 800
13 50
14 131 (41, 43, 47)
15 72

Measurement

Room	Area	Cost
A	16 m^2	$1040
B	9 m^2	$585
C	18 m^2	$1170

Measurement

Hands on.

UNIT 22 Number and Algebra

SET 1

1 5 r 4
2 290
3 6511
4 27
5 6.9
6 135
7 $1.05
8 63
9 9.6
10 7
11 0.9
12 9 r 1
13 63 m
14 96
15 30

SET 2

1 $12, $48
2 $16, $64
3 $10, $40
4 $7, $28
5 20 children
6 $24
7 7 L
8 10 trees
9 30 goals
10 $800
11 25%
12 40%
13 50%
14 20%

SET 3

1

Heptagons	1	2	3	4	5	6	7
Sides	7	14	21	28	35	42	49

2 Number of sides = Number of heptagons × 7
3 77 sides

Measurement

1 4 cm × 4 cm square
2 4 cm × 3 cm rectangle

SET 4

1 0927
2 3750 m
3 23, 29, 31
4 $2.15
5 $\frac{3}{4}$
6 540°
7 55 217
8 16
9 5 cm
10 9
11 480 cm^2
12 $224
13 Cylinder
14 $9.40
15 90°

Statistics and Probability

1 Athletics, basketball, netball, swimming
2 Basketball, soccer, swimming
3 Yes
4 Soccer (22%) Australian Rules (21%) Rugby League (20%)

UNIT 23 Number and Algebra

SET 1

1 30
2 500
3 400
4 950
5 420
6 ÷
7 ×
8 33
9 1, 2, 3, 5, 6, 10, 15, 30
10 No
11 23 260
12 297 cm
13 2920
14 4 hrs
15 20 min

SET 2

1 70, 110, 260, 350, 200
2 800, 900, 1000, 1270, 3160
3 800, 1600, 2300, 10 000, 11 800
4 6000, 15 000, 50 000, 100 000, 225 000
5 $1750
6 $2380
7 $3340

SET 3

1 3 × $4 = $12
2 5 × $5 = $25
3 6 × $1 = $6
4 8 × $2 = $16
5 12 × $3 = $36
6 10 × $5 = $50
7 5 × $7 = $35
8 8 × $9 = $72
9 9 × $4 = $36
10 3 × $5 = $15
11 5 × $10 = $50
12 20 × $10 = $200

SET 4

1 2215
2 3.19, $3\frac{1}{5}$, 3.3, $3\frac{1}{3}$
3 $300
4 $\frac{1}{5}$
5 1111
6 $33.50
7 108
8 45
9 $60
10 2027
11 6.5°C
12 6:45 am
13 1296
14 6
15 26.33 m

Statistics and Probability

1 5000
2 5500
3 8500
4 8000
5 3500

Statistics and Probability

1 True
2 True
3 True
4 True
5 True

UNIT 24 Number and Algebra

SET 1

1 28
2 Yes
3 ×
4 190
5 900
6 8
7 546
8 1, 3, 9, 27
9 $815
10 8500
11 9
12 8
13 52
14 80 000
15 450 cm
16 75

SET 2

1 52.8
2 112
3 124.2
4 71.6
5 70.56
6 206.88
7 641.1
8 1344.42
9 $1219.10
10 8.75 m
11 56.56 km^2

SET 3

1 56, 48, 40, 32, 24, 16, 8, 0
2 18, 36, 54, 72, 90, 108, 126
3 110, 111, 112, 113, 114, 115, 116
4 100, 90, 80, 70, 60, 50, 40
5 Hands on.

SET 4

1 4.22
2 0.35
3 4
4 5.4
5 $\frac{9}{100}$
6 $\frac{98}{100}$
7 12
8 70
9 8
10 236
11 70°
12 1 hr 40 min
13 2 hr 35 min
14 2 hr 35 min
15 1 hr 15 min

Number and Algebra

1 $3
2 4
3 6
4 4
5 10
6 15
7 12
8 12
9 150
10 180
11 120
12 100
13 60
14 150

Measurement

1 
2
3 
4  

UNIT 25 Number and Algebra

SET 1

1 +
2 63
3 $13
4 $\frac{7}{10}$
5 30
6 −
7 ÷
8 3
9 0.6
10 30
11 No
12 No
13 235 321
14 30 minutes
15 $26

SET 2

1 $\frac{3}{8}$
2 $\frac{7}{10}$
3 $\frac{5}{6}$
4 $\frac{3}{8}$
5 $\frac{7}{10}$
6 $\frac{4}{9}$
7 $\frac{3}{10}$
8 $\frac{5}{8}$
9 $\frac{11}{8}$ or $1\frac{3}{8}$
10 $\frac{4}{10}$
11 $\frac{7}{8}$

SET 3

1 0.35, 0.40, 0.45
2 0.39, 0.41, 0.43
3 0.90, 1.05, 1.2
4 1.52, 1.62, 1.72
5 5.25, 5.65, 6.05
6 10.31, 10.39, 10.47
7 26, 31, 36, 41, 46

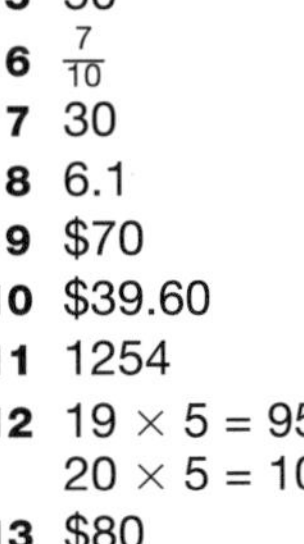

SET 4

1 52
2 199
3 120
4 115
5 50
6 $\frac{7}{10}$
7 30
8 6.1
9 $70
10 $39.60
11 1254
12 19 × 5 = 95 or 20 × 5 = 100
13 $80
14 98, 75, 52
15 1, 2, 3, 6, 7, 14, 21, 42
16 107 runs
17 36 cm

Space

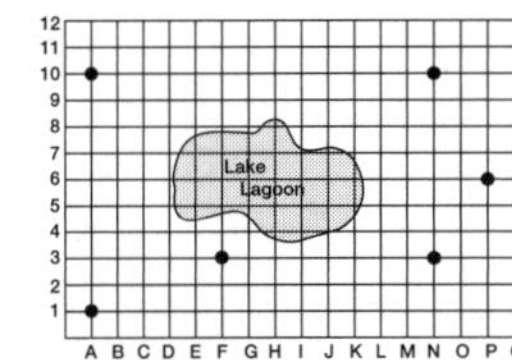

7 90 km
8 130 km
9 70 km

Space

1 Rectangle
2 Right-angled triangle

UNIT 26 Number and Algebra

SET 1

1 $5.80
2 1, 3, 5, 9, 15, 45
3 2.4
4 $\frac{9}{10}$
5 108
6 $1.20
7 13.6
8 No
9 9
10 60
11 42
12 9 r 5
13 10:06
14 19
15 9

SET 2

1 4
2 2
3 4
4 3
5 4
6 4
7 4
8 6

SET 3

1 2.12
2 2.32
3 5.1
4 9.1
5 3.2
6 2.31
7 3.64
8 1.23
9 3.42
10 4.41
11 5.31 m
12 6.15 cm

SET 4

1 60 cm
2 350 m
3 60 m
4 $119.94
5 762 min
6 21 min
7 $1\frac{4}{5}$ or $1\frac{8}{10}$
8 75 L
9 $90
10 $8.91
11 20 L
12 20%
13 8
14 2400 mL
15 945 mL

Measurement

1 45 cm
2 5 L
3 9 kg
4 3 hrs
5 1.5 m
6 4 m
7 3 t
8 5 km
9 8 L
10 4 hrs
11 1600 g
12 30 min
13 3250 m
14 405 cm
15 450 cm
16 150 min
17 5250 kg
18 48 hrs
19 325 cm
20 565 g
21 $30

Statistics and Probability

$ = $\frac{1}{10}$
% = 0.2
© = 30%
= 15%
⊞ = $\frac{1}{4}$

Answers

UNIT 27 Number and Algebra

SET 1

1 66
2 54
3 56
4 20
5 11:10 am
6 51
7 $72
8 1, 3, 7, 21
9 25
10 20
11 1050
12 100
13 28°
14 $1.22
15 $7.50

SET 2

1 3
2 7
3 9
4 2.7
5 5.4
6 34
7 5.66
8 87.6
9 543
10 765
11 432
12 12.3
13 125
14 2274

SET 3

1 $1100, $1025, $900, $859.50, $684, $656.70, $557.00, $517
2 No
3 $300
4 $80.50
5 $152.30

SET 4

1 20.3 m
2 $\frac{91}{100}$
3 9
4 100°
5 $8\frac{3}{5}$
6 50 000
7 $28.90
8 Various responses – e.g. 7 days × 24 hours × 60 minutes = number of minutes per week

Number and Algebra

1

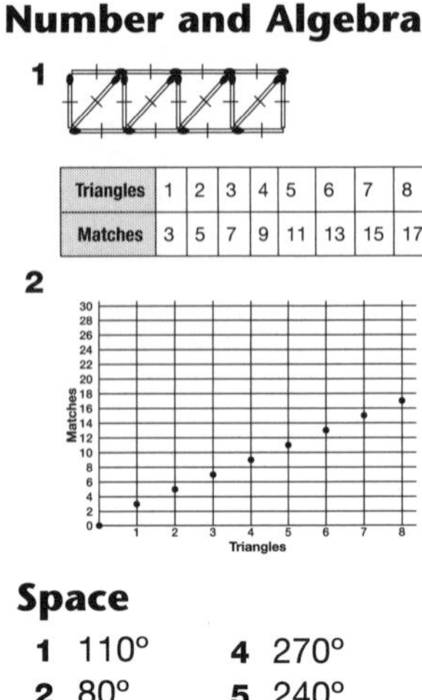

Triangles	1	2	3	4	5	6	7	8
Matches	3	5	7	9	11	13	15	17

2

Space

1 110°
2 80°
3 50°
4 270°
5 240°
6 300°

UNIT 28 Number and Algebra

SET 1

1 +
2 81
3 –
4 42
5 ÷
6 13
7 ×
8 1745
9 7526
10 $7.50
11 $7
12 8
13 15
14 6
15 35

SET 2

1 Accurate
$5 × 5 = $25
2 Inaccurate
$50 – $30 = $20
3 Accurate
5 × 10 = 50
4 Inaccurate
60 km × 3 = 180 km
5 Inaccurate
$48 ÷ 6 = $8 for one ticket
Therefore, 10 tickets would cost $80.
13 Inaccurate
1000 mL – 500 mL = 500 mL

SET 3

1 3.34
2 7.1
3 5.1
4 8.55
5 8.2
6 0.81
7 8.03
8 5.02
9 7.03
10 0.72
11 0.81
12 9.21
13

4 kg for $21.00

SET 4

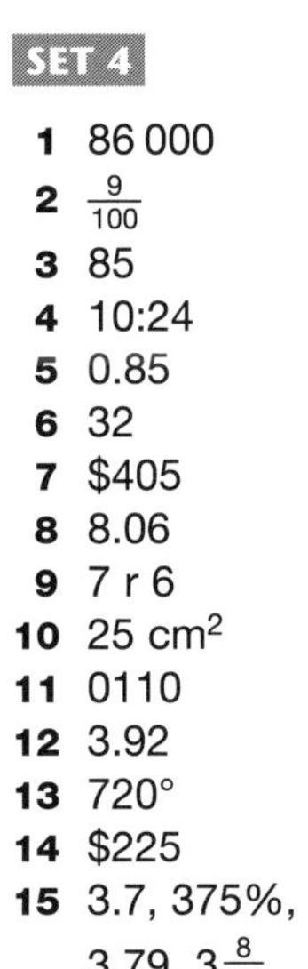

1 86 000
2 $\frac{9}{100}$
3 85
4 10:24
5 0.85
6 32
7 $405
8 8.06
9 7 r 6
10 25 cm^2
11 0110
12 3.92
13 720°
14 $225
15 3.7, 375%, 3.79, $3\frac{8}{10}$
16 $81.63

Statistics and Probability

Frequency	
Soccer	12
Cricket	8
Baseball	4
Netball	11
Rugby	6
Tennis	3
Total	44

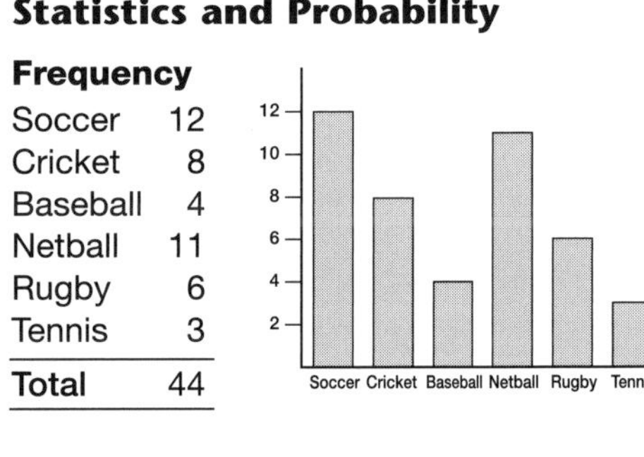

Space

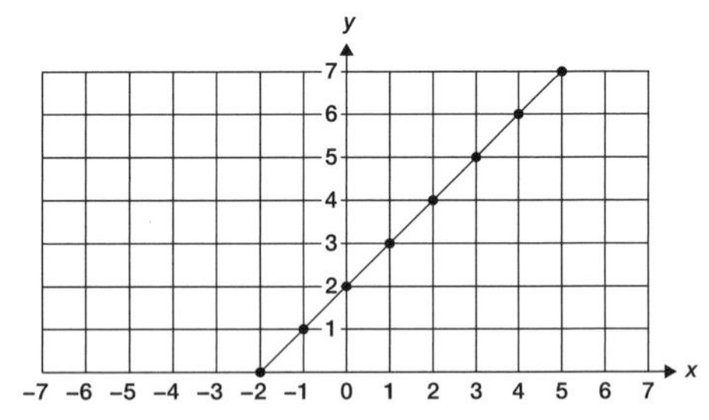

UNIT 29 Number and Algebra

SET 1

1 980
2 160
3 $16.50
4 9.6
5 6.5
6 250
7 51
8 $2.60
9 75
10 8 r 7
11 $\frac{50}{100}, \frac{5}{10}$ or $\frac{1}{2}$
12 250
13 $100
14 250
15 120

SET 2

1 ($5 × 30) + $300
2 ($3.50 × 80) + $350
3 (5000 mL – 2000 mL) – 200 mL × 10
4 (75 cm – 65 cm) × 20
5 (6 × 6) + (2 × 8)

SET 3

1 30
2 70
3 29
4 3
5 158
6 68
7 98
8 $\frac{6}{18}$ or $\frac{1}{3}$
9 12 pieces and 6 pieces

SET 4

1 11.71
2 5
3 0.09
4 5
5 Rhombus
6 321
7 12
8 32 cm
9 120°
10 5 670 000
11 18 090 000
12 3 240 000

Statistics and Probability

1 Red: 12
2 Blue: 8
3 Green: 4

Space

1 40°
2 48°
3 52°
4 28°
5 69°
6 55°
7 165°
8 30°

UNIT 30 Number and Algebra

SET 1

1 1000 mm
2 910
3 41
4 6
5 350
6 ×
7 +
8 4 r 1
9 12
10 60
11 220
12 174
13 100
14 5
15 2350 mL

SET 2

1 −3
2 −3
3 −8
4 −8
5 3
6 2
7 2
8 1
9 −2
10 1
11 −5
12 −3

SET 3

1
2
3

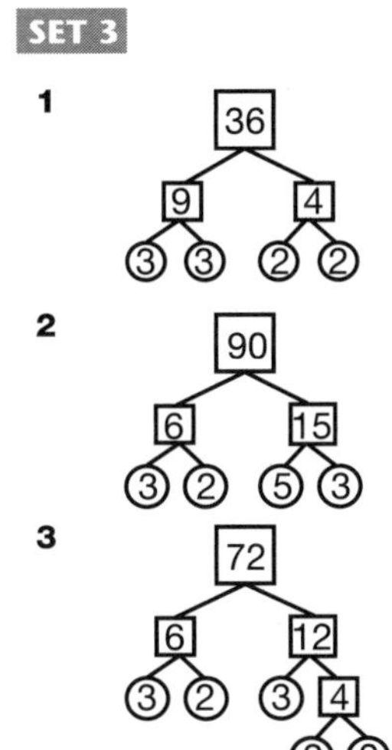

SET 4

1 585
2 $24
3 74
4 $650
5 70 × 9 = 630
6 27
7 $\frac{3}{10}$
8 7.02
9 2328
10 96 r 6
11 269 min
12 $2.70
13 90 m^3
14 2 000 000
15 2.9
16 $210

Space

1
2

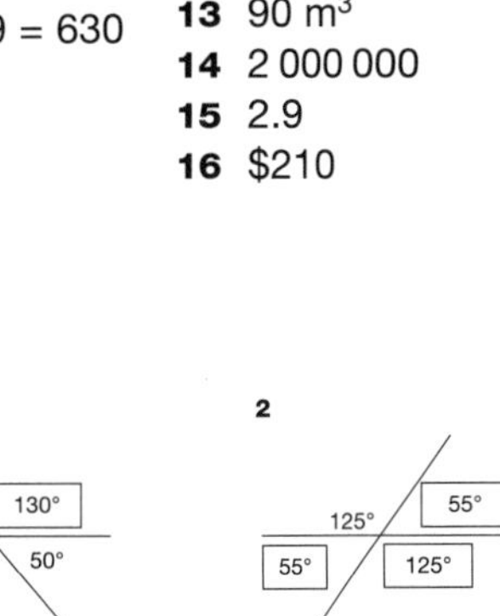

3

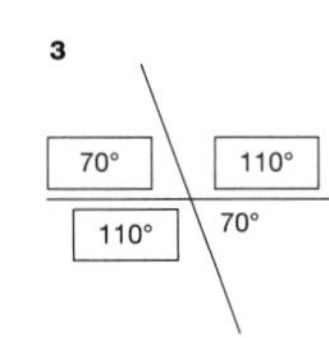

Statistics and Probability

1 False
2 True
3 True
4 False
5 True

UNIT 31 Number and Algebra

SET 1

1 410
2 ÷
3 60
4 200
5 $1.65
6 7 r 1
7 20 231
8 900 000
9 No
10 23, 29
11 1, 3, 7, 21
12 0.23
13 $\frac{1}{4}$
14 3745
15 4500

SET 2

1 11
2 9
3 15
4 21
5 25
6 25
7 23
8 21
9 21
10 22
11 0
12 −4
13 −3
14 −6
15 −28
16 −9

SET 3

1 27, 43, 59, 75, 91
2 15, 25, 35, 45, 55

SET 4

1 680
2 21
3 9.1
4 341 000
5 0.5 ha or 5000 m^2
6 6
7 49 cm^2
8 $\frac{26}{100}$
9 100 000
10 197
11 $480
12 50
13 five past four
14 93 mm
15 3.7 km
16 24

Space

Hands on.

Measurement

1 2 L
2 5 L
3 7000 mL
4 3500 mL
5 7250 mL
6 2 kL
7 8 kL
8 5000 L
9 2500 L
10 4 000 000 L

UNIT 32 Number and Algebra

SET 1

1 81
2 $1.70
3 400
4 27
5 6 r 4
6 $2800
7 $13
8 65
9 112
10 9.83
11 3.2
12 74
13 108
14 $\frac{6}{10}$
15 112

SET 2

1 $15 162
2 $7250
3 $7200
4 $4977
5 $247.50
6 $361
7 $877
8 $186.30
9 $829
10 $402.50

SET 3

1 9
2 1.2
3 6
4 1.5
5 5
6 3
7 13
8 0.5
9 42
10 8
11 ×, +
12 +, ×
13 −, ×
14 ×, −

SET 4

1 True
2 35 240
3 20
4 3
5 25 000
6 6000
7 37°C
8 5 302 000
9 225 cm^2
10 120 000
11 $82.80
12 $225
13 4350
14 4680 seconds
15 500
16 546
17 70
18 172 cm

Space

1 A = (4, 4)
2 B = (2, 3)
3 C = (1, 1)
4 D = (4, 1)
5 E = (−4, 3)
6 F = (−2, 1)
7 G = (−3, −2)
8 H = (2, −2)

Statistics and Probability

1 True
2 True
3 True
4 False
5 True
6 True

Answers

UNIT 33 Number and Algebra

SET 1

1 5 r 1
2 2 L
3 3400
4 ×
5 +
6 70
7 15
8 3 r 3
9 37
10 50
11 1500
12 \$23
13 37 726
14 \$12
15 \$45
16 5 (7, 11, 13, 17, 19)

SET 2

1 2.31, 23.1, 231
2 43.8, 438, 4380
3 6.43, 64.3, 643
4 18.7, 187, 1870
5 156,1560, 15 600
6 356.8, 35.68, 3.568
7 429.5, 42.95, 4.295
8 23.52, 2.352, 0.2352
9 6.89, 0.689, 0.0689
10 1.95, 0.195, 0.0195
11 True
12 True
13 True
14 True
15 False
16 True

SET 3

1 1.6, 2, 2.4
2 4.48, 4.64, 4.8
3 8.33, 8.44, 8.55
4 4.69, 4.92, 5.15
5 5.51, 5.68, 5.85
6 6.09, 6.12, 6.15
7 5, $5\frac{2}{3}$, $6\frac{1}{3}$
8 $6\frac{1}{4}$, 7, $7\frac{3}{4}$
9 $6\frac{1}{5}$, $6\frac{3}{5}$, 7
10 $7\frac{4}{5}$, $8\frac{2}{5}$, 9

SET 4

1 8750
2 \$42.75
3 155
4 True
5 12
6 20%
7 100
8 87 km/h
9 16.2 m
10 120
11 326
12 0.1, 0.11, 1.0, 11.1
13 \$200

Measurement

1 180 m^2
2 1080 m^3
3 \$324 000

Statistics and Probability

Totals: 35, 10, 120
1 Lane
2 Vanda
3 Yes 4 No 5 Yes

UNIT 34 Number and Algebra

SET 1

1 500
2 ÷
3 270
4 +
5 600
6 92
7 False
8 76 000
9 25
10 78
11 True
12 560
13 60
14 False
15 \$1.80

SET 2

1 $5\frac{1}{3}$
2 $6\frac{1}{4}$
3 $51\frac{1}{5}$
4 $247\frac{1}{3}$
5 $86\frac{1}{4}$
6 $59\frac{2}{5}$
7 $322\frac{2}{7}$
8 $433\frac{4}{8}$
9 $733\frac{1}{9}$
10 83.25
11 198.5
12 126.25
13 186.75
14 389.7
15 267.9
16 73.2
17 1153.4
18 1075.6

SET 3

1 21
2 11
3 105
4 125
5 25
6 19
7 40
8 45
9 72
10 21
11 12
12 74

SET 4

1 754
2 15.5
3 48
4 94 km/h
5 33%
6 0
7 \$126
8 $\frac{3}{4}$
9 120°
10 Various solutions such as
10 × 6 × 1
5 × 6 × 2
5 × 3 × 4
12 × 5 × 1
20 × 1 × 3

Space

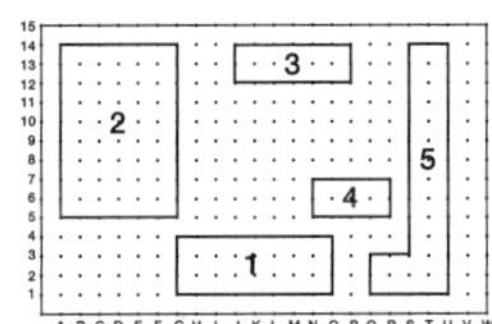

Number and Algebra

1 130.26 g
2 183.83 g
3 119.2 g
4 97.3 g

UNIT 35 Number and Algebra

SET 1

1 58
2 \$20
3 20
4 ÷
5 100
6 –
7 214
8 72
9 0.5
10 No
11 12:07
12 60 000
13 39 241
14 72
15 366
16 1461

SET 2

1 73.4
2 1216.25
3 1624.75
4 155.4
5 824.25
6 324.375
7 1870.6
8 1212.125
9 357.375
10 $131.\dot{3}$
11 $641.1\dot{6}$
12 $357.\dot{1}$
13 $257.\dot{3}$
14 $616.1\dot{6}$
15 $567.\dot{2}$
16 $2547.\dot{3}$
17 $254.8\dot{3}$
18 $487.\dot{8}$

SET 3

1 6
2 5
3 18
4 8
5 2
6 25
7 5
8 4
9 (\$500 – \$125) × 52 = \$1950

SET 4

1 True
2 216 cm
3 4
4 14 750 kg
5 35
6 120 × 7 = 840
7 32
8 2500 ÷ 5 = 500
9 60
10 True
11 2327
12 280
13 2160 min
14 $\frac{6}{100}$
15 29, 17, 41, 131, 251, 443

Statistics and Probability

Age	Tally	Frequency
14					3
15	𝍸	5
16	𝍸 ||	7
17	𝍸 𝍸 |||	13
18	𝍸 𝍸 𝍸	15
19	𝍸 ||	7

Measurement

1

Kilograms	5467 kg	6381 kg	9567 kg	4607 kg	7859 kg	3555 kg	8016 kg
Tonnes and kilograms	5 t 467 kg	6 t 381 kg	9 t 567 kg	4 t 607 kg	7 t 859 kg	3 t 555 kg	8 t 016 kg
Tonnes	5.467 t	6.381 t	9.567 t	4.607 t	7.859 t	3.555 t	8.016 t

2 3.789 t $<$ 3.897 kg 3 5.679 kg $>$ 5.6 t 4 5 t 452 kg $<$ 5.5 t 5 5900 kg $>$ 5.899 t